AI
The End of Education

Toward a New Civilization of Human Growth

ButterflyMan

ButterflyMan Publishing LLC

United States of America

First paperback edition 2026

ISBN: 979-8-90217-018-1

Published by ButterflyMan Publishing LLC
United States of America
Website: https://www.butterflyman.com
Email: contact@butterflyman.com

This book is a work of philosophical analysis. It does not provide professional, legal, or educational advice.

Preface

What Is Education, Really?

What is education, really?
It is a question humanity has repeated for centuries, yet rarely confronted with honesty.

I come from the Chinese education system. The national college entrance examination—*Gaokao*—is often praised, both inside and outside China, as a universal and fair mechanism of social mobility. Many admire it as an ultimate symbol of equality. Yet few ask where it truly comes from, why it exists, and what it is ultimately designed to serve.

In reality, the Gaokao is not rooted in concern for human development. It is a product of political necessity and social control. Its core function is stability. By elevating test scores above all else, it locks an entire generation of young people into classrooms, confining what should be the most dynamic force of social renewal inside an exam-driven cage.

The fundamental purpose of this system is not liberation, but the preservation of a rigid political and social order, ensuring that existing power structures and vested interests can reproduce themselves indefinitely. Its most deceptive feature is the illusion of equality. In a society where political and social rights are fundamentally unequal, the idea of a "fair" examination is itself a contradiction. True equality cannot exist inside a structurally unequal system.

Score-based selection allows only a tiny minority of students to "leap through the gate." These few are then used as living proof of the system's legitimacy—evidence meant to persuade millions of others to continue believing in merit through numbers. In doing so, attention is diverted from a more fundamental truth: real equality is not equality of scores, but equality of political and social rights.

3

The irony is that many of those who succeed within this system quickly forget their origins. Once elevated into elite positions, they often become its most determined defenders. Having benefited from the structure, they learn to replicate its methods— managing, filtering, and controlling others—becoming new gatekeepers themselves.

This is the essence of the Chinese education system.

By contrast, British education forms the historical foundation of modern Western education. Its emphasis on institutions, rationality, and public norms was later expanded and intensified in the United States. Supported by strong economic foundations and massive investment in research, Western higher education became a global benchmark for scientific innovation and knowledge production.

Nordic education, however, developed along a different path. Shaped by unique political cultures and human-centered values, it rejected extreme competition and returned education to a people-centered model. This approach has made the Nordic region one of the world's most important laboratories for social innovation.

Yet in the age of artificial intelligence, all traditional education systems are approaching their historical limits.

When knowledge is no longer scarce, when cognitive capacity can be universally amplified by technology, and when learning no longer depends on teachers, classrooms, or institutional hierarchies, education as a system designed to shape, rank, and manage human beings has completed its historical mission.

We are now entering a new dawn—a phase of civilization in which humanity and artificial intelligence coexist. This moment demands that we redefine equality, happiness, and the purpose of a human life. After the end of education, self-directed growth becomes the defining theme of the AI era.

Today, humanity stands between two forces: the immense power released by technological progress, and the fear and uncertainty this power generates among political and social elites. Yet I firmly believe that if AI belongs to all citizens and to humanity as a whole—rather than being monopolized by a small oligarchy—it can become a genuinely emancipatory force.

In that future, AI will not deepen inequality. Instead, it can help humanity move closer to a truly free, equal, and dignified society—one that earlier generations could only imagine as a utopia.
This is not a future we should fear.
It is a future we should consciously choose to embrace.

ButterflyMan
San Francisco
Jan 9th , 2026

Contents

Preface

Overview of Chapter Structure

Part I — A Civilizational Portrait of Global Education Systems

(The Final Round of Experiments in the Pre-AI World)

Part I: Global Education Systems — The Last Pre-AI Experiments

The core purpose of this part is **not** to judge which education system is better or worse.

It raises a question that is more brutal—and more fundamental:

**Before the emergence of AI, what kinds of methods did humanity attempt in order to "shape people"?
And what price did society pay for each method?**

This is the final systematic inventory of the **Pre-AI education world**.

Chapter 1 — The Structural Limits of Traditional Education in the Age of AI

The British Education System — Highly Mature, Yet at the End of Its Historical Limit

The British education system represents the highest level of completion that industrial-era education has ever achieved.

If even this system begins to fail in the age of AI, then the problem is not institutional management or national capacity, but the paradigm of **"education" itself**.

- **"Elite education" conceals deep structural inequality**

Family background, tutoring resources, and social class are legalized and continuously reinforced through exam systems.

- **When Britain must transform, no country is exempt**

This provides the theoretical foundation for why poor countries may be able to *skip the traditional education stage altogether*.

Chapter 2 — American Education: The Shell of Freedom, the Core of Selection

Core Logic
- Surface: freedom of choice, individual development
- Substance: extreme selection accomplished through markets and elite schools

Strengths
- World-class innovators
- Strong expressive power and cross-domain capability

Systemic Costs
- Freedom becomes intensely class-dependent
- Large populations are quietly abandoned
- Learning degenerates into investment behavior

AI Breaking Point

When knowledge and tools are no longer monopolized by elites, the "expensive ticket" of American education loses its legitimacy.

Chapter 3 — Japanese Education: Order, Collectivism, and the Success Built on Heavy Internal Exhaustion

Core Logic
- Training stable, reliable executors for industrial society

Strengths
- Extremely high discipline
- Strong social coordination and a sense of responsibility

Systemic Costs
- Individual differences are flattened
- Creativity and well-being are continuously overdrawn

AI Breaking Point

AI does not need "perfect executors."
Japanese-style education loses its historical stage.

Chapter 4 — German Education: The Rationality of Tracking, and the Risk of Early Fixation

Core Logic
- Social function takes priority over individual possibility

Strengths
- Occupational dignity
- A skilled-worker system
- Industrial stability

Systemic Costs
- Potential is judged too early
- "Late awakeners" are ignored by the system

AI Breaking Point

When skills can be rapidly reconstructed,
early tracking becomes a civilization-level misjudgment.

Chapter 5 — French Education: Rationality, Elites, and State-Sanctioned Success

Core Logic
- Selecting "rational elites" through national examinations

Strengths
- Deep critical thinking
- A philosophical tradition
- Public reason

Systemic Costs
- Entry routes are extremely narrow
- Large amounts of potential are blocked at the gate

AI Breaking Point

When "thinking ability" no longer depends on elite pathways,
the legitimacy of the French model begins to shake.

Chapter 6 — Chinese Education: The Sacrifice of Humanity Under Extreme Efficiency

Core Logic
- Social ranking and stability through unified examinations

Strengths
- Large-scale foundational competence
- Rural-to-urban mobility

Systemic Costs
- Humans are reduced to scores
- Creativity and dignity are systematically suppressed
- Happiness is postponed to "forever later"

AI Breaking Point

When memorization, drilling, and standard answers fully collapse,
Chinese education becomes an ox-cart in the age of airplanes.

Chapter 7 — Indian Education: Treating Education as the Only Lottery to Escape Destiny

Core Logic
- Extreme competition, extreme selection

Strengths
- World-class elites
- Export of technical talent

Systemic Costs
- The vast majority suffer structural failure
- Education becomes an amplifier of social anxiety

AI Breaking Point

When technology no longer rewards only a tiny number of winners,
lottery-style education collapses completely.

—

Chapter 8 — Latin America: The Ideal Exists, but Institutions Cannot Carry It

Core Logic

- Human-centered aspirations genuinely exist
- State capacity is insufficient

Strengths

- Creativity
- Emotional expression
- Social resilience

Systemic Costs

- Education cannot change destiny
- Hope and reality remain disconnected for the long term

AI Breaking Point

AI may instead become the first real opportunity
to truly lower entry barriers.

Chapter 9 — Finland and Nordic Education: A Successful Pre-AI Model, but Not the Destination

Core Judgment

Before the arrival of AI, Nordic education was humanity's closest experiment in "respecting people."

It validated three things:

- Low control ≠ chaos
- Respecting difference ≠ sacrificing quality
- Well-being ≠ social failure

But it is still Pre-AI

- Schools are still education institutions
- Teachers still serve as the guiding center
- Learning still happens before society fully begins

The Path Toward the Ultimate Turn

Nordic education proves that "less control can work,"
but the AI age forces us to keep asking:

When knowledge is no longer scarce,
do we still need "education" itself?

**Chapter 10 — The Ultimate Comparative Dual Chapter:
Africa × the Nordic World**

Africa and the Nordic world stand on the same historical axis,
yet occupy two completely different positions:
- **Africa:**
A human region that has almost never been fully shaped
by modern education systems
- **The Nordic World:**
The most successful Pre-AI region shaped by modern
human-centered education

Part I Summary

All education systems have been trying to answer the same
question:

How should a person be shaped in order to adapt to society?

But the AI age, for the first time, allows us to reverse the
question:

Should society stop shaping people,
and instead make room for human unfolding?

Part II — The Manifesto of Human Growth in the Age of AI

The Manifesto of Human Growth in the Age of AI

Prologue — Why We Must End "Education"

Chapter 11 — How a Human Life Unfolds: A Continuum from Childhood to Old Age

The industrial age cut life into separated stages:
- Childhood: shaped
- Youth: selected
- Adulthood: used
- Old age: exited

This division does not come from life itself,
but from the needs of system management.

Chapter 12 — Education Enters the Museum of History

Every real leap in human civilization
is accompanied by the disappearance of a word.

Not because the word was wrong,
but because it completed its mission.

Today we must admit one fact:

The word "education" no longer belongs to the future.

Chapter 13 — After the End of Education: A New Human-Centered Civilization

When "education" completes its historical mission,
the world does not fall into emptiness.

On the contrary—
a civilizational core long concealed appears for the first time.

Chapter 14 — The Fundamental Rupture Brought by AI

AI does not bring a tool upgrade;
it brings a fundamental rupture in the meaning of learning:

- Knowledge is no longer scarce
- Memory is no longer important
- Skills can be instantly assisted
- Correct answers are everywhere

What remains irreplaceable is only the human being:

- Judgment
- A sense of value
- Imagination
- Empathy
- Directional choice
- Care for others and for the world

Continuing to define people by exam scores
is a systemic wasting of humanity.

Chapter 15 — The True Role of AI: Not Teacher, but an Amplifier of Human Potential

When AI enters learning,
the most common mistake people make is this:

Treating AI as a stronger Teacher.

Chapter 16 — Redefining a Happy Life

We must answer directly a question long avoided:

What is a happy life?

Happiness is not wealth ranking,
not social reputation,
not winning by comparison.

True happiness is:

When an individual pushes their unique potential to its fullest, the inner fulfillment, intensity of life, and sense of existence that arise.

Humans are born different.
Difference is not the problem.
Forced uniformity is the tragedy.

Chapter 17 — Abolish Evaluation: Why Humans Should Not Be Measured

If exams are the most visible violence of the old education system,
then evaluation is the most hidden—and most persistent—control mechanism.

Scores, grades, rankings, GPAs, comments, labels—
they look gentle,
yet silently transform "human beings" into "comparable objects."

Chapter 18 — End Exams: A Non-Negotiable Precondition

In the AI age:
- Exams no longer measure ability
- Exams measure only obedience
- Exams manufacture fear
- Exams suppress potential

All standardized exams must be abolished.

Not reformed.
Not reduced.
But completely terminated.

Because any system that runs on exams
remains, in essence, anti-human.

Chapter 19 — Abolish Grades, Classes, and Age Order

Human growth is not linear.
Human awakening has no unified timetable.

Therefore:
- No grades
- No classes
- No age stratification

Only one form remains: **Open Sessions**
- Any age may participate
- Anyone may freely enter
- Choice based on interest, condition, and rhythm

This is institutional respect for differences in life rhythm.

Chapter 20 — The New Role of Schools: Fields of Human Potential Interaction

Schools are no longer education institutions.

They will transform into:

Public infrastructure that exists for the unfolding of human potential.

What they provide is not answers, but conditions:
- Space
- Tools
- Experimental environments
- Social interfaces
- Intergenerational exchange
- AI collaboration systems

School and society are no longer separated.
Society becomes the classroom.
Real problems become the source of learning.

Chapter 21 — The End of Teacher, and the Birth of Guide

The role of Teacher is built on a premise:

"I know; you don't."

That premise has been completely destroyed by AI.

What replaces it is the **Guide**:
- No indoctrination
- No judgment
- No control

Instead:
- Spark questions
- Provide pathways
- Identify potential
- Accompany exploration
- Guard freedom

The guide's duty is not to shape people,
but to prevent people from being flattened by the system.

Chapter 22 — Please Stop Educating Your Children

Almost all parents say one sentence:

"Everything I do is for my child's own good."

It is precisely this sentence
that forms the most stable—
and most dangerous—emotional foundation of the old education
system.

Chapter 23 — A New Definition of Learning (and the End of the Old Definition)

Learning is no longer the input of knowledge.

It is:

A process in which a person continuously unfolds themselves, understands the world, builds relationships, and assumes responsibility.

It is lifelong,
non-competitive,
unrankable,
and unquantifiable.

And precisely because of this,
it is truly important.

Chapter 24 — The New Word After "Education": Human Growth Center

We refuse to continue using the word "education."

Because:
- It assumes humans are objects to be shaped
- It implies authority and obedience
- It belongs to industrial civilization

We propose a new core concept:

What is a **Human Growth Center**?

Humans are not cultivated.
They are unfolded.

Schools do not cultivate people.

They are only responsible for not obstructing human unfolding.

Epilogue — The First Global Equality of Human Unfolding in the Age of AI

Appendix
- **The Human Growth Manifesto**

Beyond education, toward a new human society
- **Case Study**

Why poor countries should skip traditional education and directly enter the Human Growth Center era
- **Global Guide Selection and Training system**

Global Guide Framework for Human Growth Center
Global Charter for Guides of Human growth

Part I — A Civilizational Portrait of Global Education Systems

(The Final Round of Experiments in the Pre-AI World)

Chapter 1 · The Structural Limits of Traditional Education in the Age of AI

The British Education System — Highly Mature, Yet at the Upper Limit of Its Era

I. Why Must We Examine the British Education System?

Among all education systems worldwide, the British education system is often regarded as:
- The most disciplined
- The most clearly structured academically
- The most successful in elite formation
- The most globally recognized in terms of credentials

It almost represents the **absolute ceiling** of what an industrial-era education system can achieve.

Therefore, if even the British system reveals fundamental failure in the age of AI, then the problem does not lie in any particular nation, culture, or management capacity, but in one thing:

"Education" as an institutional form has itself become obsolete.

II. The Basic Structure of the British Education System

The core logic of the British education system is:

Early tracking + standardized examinations + concentrated elite cultivation

Its typical pathway includes:
- **Key Stages** (foundational phases)

- **GCSE (ages 14–16):** unified multi-subject examinations
- **A-Level (ages 16–18):** highly early specialization (usually three subjects)
- **University system:** elites heavily concentrated in institutions such as Oxford and Cambridge
- **Quality oversight:** unified evaluation by Ofsted

The symbolic apex institutions include the **University of Oxford** and the **University of Cambridge**.

This system has indeed produced, over a long period:
- Strong logical reasoning 能力
- High academic discipline
- A clear mechanism of social selection

However, the fundamental problem is this:

It was designed for a pre-AI society.

III. Why Was It Successful in the Past?

Before the emergence of AI, the British model was highly "rational":
- Knowledge was scarce → teachers functioned as authorities
- Careers were stable → early tracking was effective
- Information was expensive → examinations were necessary filters
- Social structures were fixed → credentials determined class position

In that era:
- *What* you learned mattered more than *how* you learned
- *Who* was selected mattered more than *who* grew

25

The core mission of education was **to filter people**, not **to cultivate human beings**.

IV. Structural Failure in the Age of AI

1 Premature Fixation Compresses Human Potential

British students must lock in their direction through A-Levels at the age of **16**.

But what the AI era truly requires is:
- Cross-domain capability
- Cognitive flexibility
- Nonlinear growth
- Late-blooming innovators

> **Human development is nonlinear, while the British system is rigid.**

2 Examination Signals ≠ Real Capability

GCSEs and A-Levels essentially measure:
- Test-taking technique
- Memory capacity
- Matching of standard answers

AI already outperforms humans comprehensively in these areas.

What remains irreplaceable is:
- Independent judgment
- Moral reasoning
- Creative synthesis
- Self-directed learning ability

These capacities are **almost impossible to measure through examinations**.

26

3 The Teacher Bottleneck Is Completely Broken by AI

Even the best teachers are constrained by:
- Time
- Energy
- Class size
- Unified pacing

AI, by contrast, offers:
- Unlimited patience
- Complete personalization
- Instant feedback
- Synchronization with the world's best knowledge

The British system is **teacher-centered**.
The AI era requires a **learner-growth-centered** architecture.

V. Real Inequality Hidden Beneath the "Elite Myth"

British education is often described as a "fair elite system," but in reality:
- The private tutoring industry is highly developed
- Family background exerts enormous influence
- Exam preparation itself is a form of resource competition

AI platforms, for the first time, make the following technically possible:
- The same intelligent tutor for rich and poor
- The same depth of knowledge
- The same optimization of learning pace

True educational equality becomes, for the first time, an engineering problem—rather than a moral slogan—in the age of AI.

VI. Why "Reform" Is No Longer Enough

British education cannot be repaired by simply "adding some AI."

Because all of its foundational assumptions have collapsed:

Old Assumptions	AI-Era Reality
Knowledge is scarce	Knowledge is abundant
People must be filtered	People must be released
Learning paths are fixed	Growth paths are open
Credentials define value	Capability evolves continuously

VII. From British Education to the Human Growth Center

The **Human Growth Center** does not reject the discipline of British education.
It **liberates that discipline from the cage of the industrial era**.

Traditional British Education	Human Growth Center
Age-based grouping	Capability-based progression
Exams and certificates	Real growth trajectories
Early specialization lock-in	Open exploration
Scarce teachers	AI with infinite amplification
National curriculum	Global knowledge commons

VIII. Conclusion: When Britain Must Transform, All Nations Must Transform

The British education system represents the **highest level of completion traditional education can reach**.

If even it:
- Begins to fail in the face of AI
- Cannot fully release human potential
- Cannot achieve genuine educational equality

28

Then there is only one conclusion:

"Education" as an institution has completed its historical mission.

The next stage is a new social structure centered on **human growth**.

And this explains why **poor countries may, in fact, skip traditional education entirely and move directly into the era of the Human Growth Center**.

Chapter 2 · American Education: The Shell of Freedom, the Core of Selection

I. The Promise of Freedom

Among all global education systems, American education is the one most frequently associated with **freedom**.

Freedom of choice.
Freedom of expression.
Freedom of identity.
Freedom to fail—and to start again.

On the surface, the American education system appears radically different from rigid, exam-centered models. Students are encouraged to:
- Choose their own courses
- Change majors
- Explore interests
- Express opinions
- Challenge authority

This image has made American education globally attractive, especially to societies emerging from authoritarian or exam-dominated traditions.

But this freedom is **not the core mechanism** of the system.
It is the **interface**.

II. Core Logic: Freedom as Interface, Selection as Engine

Surface Logic: Choice and Individual Development

At the visible level, American education emphasizes:
- Individual preference

- Personal interest
- Self-designed pathways
- Holistic development

Students appear to move freely within a vast landscape of options: majors, minors, electives, extracurriculars, internships, exchanges, and research opportunities.

This creates the impression that **anyone can become anything**.

Structural Logic: Extreme Selection Through Market and Prestige

Beneath this surface lies the system's true engine:

> **Extreme selection, executed through markets and elite institutions.**

> Selection does not occur primarily through national exams.
> It occurs through:
- Tuition pricing
- School district segregation
- University prestige hierarchies
- Internship access
- Network accumulation
- Brand signaling

The American system does not say *"you are eliminated."*
It simply **prices you out**, **filters you out**, or **renders you invisible**.

Selection is continuous, implicit, and rarely named.

III. Where the American System Truly Succeeds

Despite its costs, American education has achieved outcomes unmatched by most systems.

1. Production of Top-Tier Innovators

The American system excels at producing:
- Technological pioneers
- Startup founders
- Interdisciplinary thinkers
- Cultural disruptors

Its flexibility allows rare combinations to emerge: engineering + design, science + entrepreneurship, technology + storytelling.

This has been critical in fields such as:
- Silicon Valley technology
- Biotechnology
- Creative industries
- Platform economics

2. Strong Expression and Cross-Domain Capacity

American education places exceptional emphasis on:
- Writing
- Presentation
- Debate
- Narrative construction

Students are trained not only to *know*, but to *persuade*, *frame*, and *connect ideas across domains*.

This produces individuals who are:
- Confident in public discourse
- Adaptable across industries

- Capable of operating in ambiguous environments

These traits were decisive advantages in the late 20th and early 21st centuries.

IV. The Systemic Costs of American Freedom

The same structure that enables excellence also generates deep, structural damage.

1. Freedom Becomes Intensely Class-Based

American educational freedom is **not evenly distributed**.

In practice:
- Wealth buys better school districts
- Wealth buys test preparation
- Wealth buys extracurricular profiles
- Wealth buys time to explore without risk

For the upper classes, freedom is expansive.
For the working and lower classes, freedom is narrow, risky, and often illusory.

What is presented as *choice* is, in reality, **a privilege gradient**.

2. The Quiet Abandonment of the Majority

Unlike exam-based systems that fail people visibly,
the American system fails people **silently**.

Millions are not "rejected."
They are simply:
- Loaded with debt

- Channeled into low-return credentials
- Excluded from elite networks
- Left without meaningful re-entry paths

They disappear from the narrative of success.

This creates a large population of people who were **never officially told they failed**, yet live with permanent downward constraints.

3. Learning Degrades into Investment Behavior

Under market logic, education becomes:
- A financial bet
- A return-on-investment calculation
- A credential accumulation strategy

Questions shift from:
- *Who am I becoming?*

to:
- *Will this major pay off?*

Curiosity is replaced by risk management.
Learning becomes a portfolio decision.

This fundamentally distorts the human purpose of education.

V. The AI Breaking Point

The American education model depends on one critical assumption:

> **Elite knowledge, elite tools, and elite opportunity are scarce.**

AI destroys this assumption.

1. Knowledge Is No Longer Elite-Owned

AI systems now provide:
- Advanced explanations
- Personalized tutoring
- Coding assistance
- Research synthesis

At near-zero marginal cost.

The informational advantage once exclusive to elite universities is rapidly eroding.

2. Tools Are No Longer Gatekept

Design tools, programming environments, analytics, writing assistants, and research platforms are increasingly accessible to anyone with connectivity.

What once required:
- Top professors
- Exclusive labs
- Elite institutional access

Can now be partially replicated or augmented outside the system.

3. The "High Ticket Price" Loses Legitimacy

When:
- Knowledge is abundant
- Tools are accessible
- Skills can be rapidly acquired

The justification for:
- Extreme tuition
- Credential inflation

- Network monopolies

begins to collapse.

The American education system's price no longer corresponds to its unique value.

VI. Structural Contradiction in the AI Era

American education still assumes:
- Scarcity-based advantage
- Winner-take-all outcomes
- Narrow elite absorption

But AI economies increasingly require:
- Broad human capability
- Distributed creativity
- Lifelong adaptability
- Late and repeated reinvention

A system optimized for **selecting a few** becomes dysfunctional in a world that needs **everyone to grow**.

VII. Why Reform Is Insufficient

Adding AI tools to American education does not resolve the contradiction.

Because the problem is not **technology**, but **architecture**.

As long as education remains:
- Market-priced
- Prestige-ranked
- Credential-centered

AI will simply **amplify inequality**, not eliminate it.

VIII. Conclusion: Freedom Without Growth Is Not Freedom

American education offered humanity a crucial experiment:

What if we replaced obedience with choice?

That experiment succeeded—partially.

But it stopped short.

Freedom without structural support for human growth becomes:
- Anxiety for the many
- Accumulation for the few
- Waste of collective potential

In the AI era, the question is no longer:

How do we select the best?

But:

How do we ensure that every human being can continue to grow?

The American model cannot answer this question within its existing framework.

This makes the transition toward a **Human Growth–centered system** not optional—but inevitable.

Chapter 3 · Japanese Education: Order, Collectivism, and Success Built on Extreme Internal Consumption

I. Why Japan Must Be Treated as a Civilizational Case

If British education represents the **upper limit of industrial rationality**,
and American education represents an **experiment in freedom under market selection**,
then Japanese education represents a third and very different extreme:

> **The completion of social order through education.**

> Japan is not chaotic.
> Japan is not inefficient.
> Japan is not underdeveloped.

> On the contrary, Japan is one of the most **disciplined, coordinated, and socially stable societies** modern civilization has ever produced.

> And the central mechanism that made this possible is education.

> This makes Japan an unavoidable case study:

> **What happens when education succeeds completely at producing order?**

II. Core Logic: Training Reliable Executors for an Industrial Society

The foundational logic of Japanese education can be summarized in one sentence:

> **Collective stability takes precedence over individual difference.**

The system is not designed to cultivate uniqueness.
It is designed to produce individuals who are:
- Predictable
- Reliable
- Cooperative
- Endurance-oriented
- Capable of functioning continuously within a system

This logic is perfectly aligned with the core demands of an industrial economy:
- Process discipline
- Precise execution
- Team synchronization
- Error minimization
- Long-term organizational loyalty

Japanese education does not ask, *"Who are you?"*
It asks, *"Can you function without disrupting the whole?"*

—

III. Structural Characteristics of the Japanese Education System

Japanese education is defined by a set of highly stable and mutually reinforcing structures:
- Nationally standardized curricula
- Behavioral training beginning in early childhood

- Collective responsibility mechanisms (cleaning duties, group discipline, shared tasks)
- High-pressure examinations at key transition points
- Tight coupling between schools, corporations, and society

From a very young age, students are systematically trained to:
- Suppress personal emotion
- Avoid standing out
- Read the social atmosphere
- Endure pressure silently

Within this system, **success is not originality**, but **error-free execution within expectations**.

IV. Where Japanese Education Truly Succeeds

It is impossible to deny the achievements of Japanese education.

1 Extreme Discipline and Reliability

Japanese students consistently demonstrate:
- Exceptional concentration
- Deep respect for process
- Strong responsibility for collective outcomes
- Very low tolerance for negligence or error

These traits translate directly into:
- Manufacturing excellence
- Infrastructure reliability
- Corporate coordination
- High levels of social trust and safety

From a system-stability perspective, this is close to a perfect model.

2 Social Cooperation and Moral Cohesion

Japanese education successfully internalizes:
- Consideration for others
- Respect for shared space
- Sensitivity to group harmony
- Willingness to sacrifice individual comfort for the collective

As a result, Japanese society exhibits remarkable resilience under external shocks.

This is the peak form of an order-based civilization.

V. The Systemic Cost: Continuous Consumption of the Individual

The very success of this system generates deep, long-term, and cumulative costs.

1 Individual Difference Is Systematically Flattened

From an early age, "difference" is treated as risk.

Those who are:
- Highly imaginative
- Emotionally sensitive
- Nonlinear thinkers
- Late bloomers

are often labeled as:
- "Problematic"
- "Uncooperative"
- "Immature"

The system does not adapt to the individual.

The individual must adapt to the system—or be marginalized.

2 Creativity Is Structurally Depleted

Japanese education produces outstanding executors, but structurally suppresses:
- Risk-taking
- Open disagreement
- Trial-and-error
- Radical reimagining

As a result:
- Innovation becomes incremental
- Disruption is imported rather than generated
- Younger generations are extremely cautious about initiating change

This is one reason Japan has struggled to produce new global platform-level innovations.

3 Happiness Is Deferred Indefinitely

Perhaps the most severe cost is psychological.

The system continuously teaches individuals:
- Endure now
- Succeed later
- Live after retirement

But the "later" rarely arrives.

The result is:
- Chronic high pressure
- Emotional suppression
- Social withdrawal
- Burnout and hollowing-out

The system functions well precisely because people are exhausted.

VI. The AI Breaking Point: When Strength Becomes Liability

The AI era introduces a fundamental inversion:

AI does not need "perfect executors."

What AI excels at is exactly what Japanese education has long optimized:
- Precision
- Repetition
- Obedience
- Optimization

This produces a brutal reversal:

The abilities Japanese education perfected are the first to be automated.

1 Execution Is No Longer a Human Advantage

In an AI-driven world:
- Accuracy is cheap
- Obedience is programmable
- Endurance is machine-native

Human value shifts toward:
- Question-forming ability
- Ethical judgment
- Imagination
- Meaning-making

43

These are precisely the capacities Japanese education has historically underdeveloped or suppressed.

2 Collectivism Without Creativity Becomes Fragile

When environments change rapidly,
systems optimized for stability struggle to pivot.

Japan's educational success turns into a structural weakness:
- Risk aversion slows transformation
- Conformity delays response
- Fear of disruption suppresses innovation

In the AI era, **adaptability outperforms discipline**.

VII. Why "Reform" Cannot Save the Japanese Model

Japan cannot resolve this crisis by:
- Adding creativity courses
- Expanding coding education
- Promoting entrepreneurship slogans

Because the problem is not curriculum.
It is **civilizational logic**.

As long as education is designed to:
- Minimize deviation
- Preserve harmony
- Maintain predictability

it will remain incompatible with an era that rewards:
- Divergence
- Exploration
- Late awakening
- Continuous reinvention

VIII. From Japanese Education to the Human Growth Center

The Human Growth Center does not reject Japan's strengths.

It **preserves order without self-destruction**.

Japanese Traditional Education	Human Growth Center
Uniform progression	Individual growth rhythm
Behavioral conformity	Internal autonomy
Endurance as virtue	Well-being as foundation
Suppression of difference	Protection of difference
Human as executor	Human as meaning-maker

Discipline is no longer enforced **against the self**,
but cultivated **in service of growth**.

IX. Conclusion: Order Without Growth Is Unsustainable

Japanese education answered one historical question with extraordinary success:

How do we build a society that functions smoothly?

But the AI era asks a different question:

When machines execute perfectly, how do humans continue to grow?

A system designed for execution cannot answer a question about growth.

Japanese education did not fail.
It **completed its historical mission**.

What comes next cannot be adjustment or repair.
It must be transformation:

From education that consumes people

45

to social structures that allow humans to unfold.

This is why, despite its success,
Japanese education cannot survive the AI era in its existing
form.

Chapter 4 · German Education: The Rationality of Tracking, and the Risk of Premature Fixation

I. Why the German Education Model Matters

Among global education systems, the German model is often regarded as the most **structurally rational**.

It is neither driven by examination obsession (as in China), nor by market competition disguised as freedom (as in the United States).
Instead, German education is built on a clear social philosophy:

> **Education exists primarily to serve social function.**
>
> Its goal is not to maximize individual possibility, but to ensure that society functions smoothly, predictably, and sustainably.
>
> For more than a century, this model has been widely admired—especially by policymakers—for its ability to produce:
> *   Skilled workers with dignity
> *   Stable industrial capacity
> *   Low youth unemployment
> *   Strong alignment between education and labor
markets

Yet it is precisely this strength that becomes its deepest vulnerability in the age of AI.

II. Core Logic: Social Function Over Individual Possibility

The foundational logic of the German education system can be summarized in one sentence:

Society's needs take precedence over individual potential.

This philosophy is not accidental. It is rooted in Germany's historical experience:
* Late industrialization
* Strong manufacturing tradition
* Social market economy
* Deep distrust of speculative instability

Education is therefore designed not to ask:

"What could this individual become?"

But rather:

"Where does this individual best fit within the social system?"

III. The Structure of German Educational Tracking

Germany's education system is defined by **early differentiation**, usually beginning around the age of **10–12**.

Typical pathways include:
* **Gymnasium** → academic track → university
* **Realschule** → applied middle track
* **Hauptschule** → vocational track
* **Dual System (Duales System)** → apprenticeship + industry integration

By early adolescence, students are effectively sorted into **life trajectories**.

This system is often defended as:
- Realistic
- Efficient
- Honest about differences
- Protective of non-academic dignity

And historically, it was.

IV. What Germany Did Right

Any serious analysis must acknowledge what the German system achieved exceptionally well.

1 Restoration of Occupational Dignity

Unlike societies where manual or technical work is stigmatized, Germany institutionalized **equal dignity of professions**.

A master craftsman (*Meister*) is not socially inferior to a university graduate.
Skill is not treated as failure; it is treated as contribution.

This prevented:
- Mass academic inflation
- Credential vanity
- Social contempt for technical labor

In this regard, Germany solved a problem that many countries still struggle with.

2 One of the World's Strongest Skilled-Worker Systems

Germany's **dual vocational system** remains one of the most successful workforce training models ever created.

It ensured:
- Seamless school-to-work transitions

- High employability
- Deep industry competence
- Intergenerational skill transmission

This directly supported:
- Automotive dominance
- Precision manufacturing
- Engineering excellence
- Export stability

Few education systems have been so tightly aligned with productive reality.

3 Long-Term Industrial Stability

Because education and labor were synchronized, Germany avoided:
- Massive youth unemployment
- Social alienation of non-elites
- Sudden workforce mismatches

For an industrial economy, this was an extraordinary achievement.

V. The Hidden Cost: Premature Judgment of Human Potential

Yet the same structure that created stability also imposed a silent cost.

1 Human Potential Is Assessed Too Early

Tracking decisions are often made when children are:
- Cognitively unfinished
- Emotionally undeveloped
- Highly sensitive to environment
- Strongly influenced by family background

At that age, what is being measured is often:
- Early compliance
- Linguistic advantage
- Middle-class cultural alignment

—not long-term creative or cognitive capacity.

In other words:

The system mistakes early performance for lifelong potential.

2 Late Bloomers Are Structurally Invisible

Human development is not linear.

Some individuals:
- Mature cognitively later
- Discover motivation only after exposure
- Excel through cross-domain synthesis
- Flourish after failure

The German system has little tolerance for this pattern.

Once tracked downwards, mobility exists in theory—but in practice is rare, costly, and socially discouraging.

As a result:
- Late awakeners are overlooked
- Reinvention is institutionally difficult
- Potential remains dormant, not because of lack of ability, but because of timing

VI. The AI-Era Breakpoint

The German system rests on a crucial assumption:

Skills are stable, scarce, and slow to acquire.

AI destroys this assumption.

1 Skills Can Now Be Rapidly Reconstructed

With AI:
- Technical skills can be learned faster
- Knowledge is no longer gatekept
- Tool proficiency can be augmented instantly
- Retraining becomes continuous

The cost of being "on the wrong track" decreases technically—but increases psychologically and institutionally if the system remains rigid.

2 The Future Requires Reconfiguration, Not Fit

The AI era demands:
- Cross-domain thinking
- Repeated reinvention
- Late-stage transformation
- Human-AI collaboration skills

These are precisely the capacities **least compatible** with early, rigid sorting.

What once was rational allocation now becomes:

A civilizational misclassification error.

VII. Why Reform Is Insufficient

Some propose:
- Adding digital skills
- Enhancing vocational flexibility
- Improving permeability between tracks

But these are surface corrections.

As long as the core logic remains:
- Early judgment
- Fixed trajectories
- Social optimization over human unfolding

AI will expose and amplify the system's limitations.

VIII. From Tracking Society to Growth Society

The question Germany now faces is not technical, but philosophical:

> **Should education allocate people to society—or allow people to unfold and reshape society?**

> In the AI era, stability no longer comes from fixed roles, but from **adaptive capacity**.

> A Human Growth–oriented system would replace:
> - Early tracking → continuous capability mapping
> - Fixed professions → evolving skill constellations
> - One-time decisions → lifelong recalibration

IX. Conclusion: When Rationality Becomes a Trap

German education represents the **most rational expression of industrial-era logic**.

And that is precisely why it now stands at risk.

When:
- Skills are no longer fixed
- Productivity is no longer linear
- Human value is no longer role-based

53

Then early tracking ceases to be rational.

It becomes:

A systemic underestimation of humanity itself.

The AI era does not punish inefficiency—it punishes rigidity.

And any system that decides **too early who a human being is allowed to become** will inevitably fail, not because it was foolish, but because it was **too rational for a world that has fundamentally changed**.

Chapter 5 | French Education: Rationality, Elites, and State-Sanctioned Success

I. Why the French Education System Must Be Examined

If German education represents the **rational allocation logic of industrial society,**
then French education represents something older, more philosophical, and more self-conscious:

> **The selection of a small number of individuals through reason, formally authorized by the state to embody public rationality.**

Among global education systems, the French model is unique.

It does not seek mass success.
It does not prioritize market efficiency.
It does not disguise hierarchy behind "choice" or "freedom."

Instead, it openly asserts a foundational belief:

A republic requires a small number of highly trained rational elites to think, judge, and govern on behalf of all.

French education was never designed to produce the largest possible pool of employable talent.
It was designed to produce **the smallest possible number of individuals with maximum intellectual legitimacy**.

In the pre-AI world, this model functioned remarkably well.

In the AI era, its legitimacy is fundamentally challenged.

II. Core Logic: Selecting "Rational Elites" Through State Examinations

The core logic of French education can be summarized in one sentence:

The state uses extreme, standardized examinations to select a minority capable of abstract rational thought.

This rationality is not primarily technical or operational.
It is a specific cognitive style, including:

- Abstract reasoning
- Conceptual construction
- Logical argumentation
- Analytical writing
- The ability to reason from universal principles

France does not conceal this aim.
On the contrary, it treats this selection as a cornerstone of the Republic itself.

Education, in this system, is not neutral—it is **a political instrument of rational legitimacy**.

III. Structural Characteristics of the French Education System

The French system is highly centralized and standardized.
Its most critical gateways include:

- **The Baccalauréat**: a national examination, with philosophy as a core subject

- **Classes préparatoires**: ultra-intensive preparatory classes with extreme attrition
- **Grandes Écoles**: elite institutions such as École Normale Supérieure, École Polytechnique, and the former ENA
- **State-recognized elite pathways**: direct pipelines into public administration, academia, and national leadership

The defining feature of this system is clear:

It does not cultivate diverse forms of intelligence. It purifies and concentrates one form: abstract rationality.

IV. Where French Education Truly Succeeds

Any serious analysis must acknowledge that French education achieves world-class outcomes in several key dimensions.

🔢 Systematic Training in Deep Critical Thinking

France is virtually **the only country in the world** where philosophy is a mandatory subject for all students at the national examination level.

Students are rigorously trained to:
- Construct concepts
- Analyze propositions
- Identify hidden assumptions
- Build logical arguments
- Separate facts from value judgments

As a result, French intellectual elites excel in:
- Theoretical reasoning
- Legal interpretation
- Policy design

- Public debate and discourse

This is not accidental—it is engineered.

2 Philosophical Continuity and Intellectual Coherence

French education is deeply embedded in a continuous intellectual tradition:
- Cartesian rationalism
- Enlightenment philosophy
- Republican universalism
- Secular public reason

This continuity gives French society a remarkable capacity for:
- Conceptual consistency
- Shared political language
- Principle-based public argument

Few societies preserve this level of philosophical coherence at a national scale.

3 State Institutionalization of Public Rationality

French education successfully binds **thinking ability to state legitimacy**.

Through its elite pathways, the system continuously reproduces:
- Judges
- Senior civil servants
- Policy architects
- Academic authorities

These individuals are not merely administrators.
They function as **symbolic carriers of republican rationality**.

In this sense, education is inseparable from governance.

V. The Hidden Systemic Costs Behind the Aura of Rationality

Yet the very structure that produces this success also generates deep exclusion.

1 Extremely Narrow Entry Points

The reality of French elite education is harsh:
- Extreme competition in preparatory classes
- Very high psychological and physical attrition
- Highly concentrated success pathways
- Strong dependence on family cultural capital

As a result:

Rationality is not equally accessible—it is accessible only to those who can survive the selection.

2 Massive Exclusion of Alternative Human Potential

The system strongly favors:
- Abstract reasoning
- Extended writing
- Sustained intellectual endurance

It systematically marginalizes:
- Interdisciplinary thinkers
- Creative integrators
- Emotional intelligence
- Experimental and practice-oriented learners
- Late bloomers

These individuals are not incapable of thought.
They simply **do not fit the state-defined template of
"legitimate rationality."**

3 Elite Legitimacy Depends on Exclusion

A rarely stated but fundamental premise of the French model is
this:

> **Elites derive legitimacy precisely because most
> people are excluded.**

> Once the exclusivity weakens,
> the symbolic authority of the entire system begins to
> erode.

VI. The AI-Era Fracture Point

French education rests on a core assumption:

> **Thinking ability must be acquired through
> scarce, elite pathways.**

> AI is dismantling this assumption at its foundation.

1 Abstract Reasoning Is No Longer Elite-Exclusive

AI systems can already:
- Assist logical analysis
- Support philosophical writing
- Structure arguments
- Provide interdisciplinary perspectives

These capacities are rapidly escaping elite institutional
boundaries.

2 The Scarcity of "Thinking Training" Disappears

When:
- Conceptual tools become widely available
- Writing and analysis are cognitively amplified
- Background knowledge is instantly accessible

The premise that "only a few can think" collapses.

3 The State Is No Longer the Sole Authorizer of Rationality

With AI and global knowledge platforms:
- Public discourse becomes decentralized
- Intellectual production becomes distributed
- Legitimacy emerges from networks, not institutions

The closed loop of **state → elite education → public authority** begins to break.

VII. Why Reform Is Insufficient

Adding:
- Technical skills
- Digital courses
- AI tools
- Broader access

cannot resolve the core contradiction.

Because the system still relies on:
- Scarcity
- Symbolic exclusion
- State-monopolized rational legitimacy

AI demands the opposite:

- Universal access to reasoning tools
- Distributed rationality
- Multiple forms of legitimate judgment

VIII. From "Rational Elites" to Universal Growth Rationality

The true challenge facing France is not educational reform, but **a transformation of what rationality itself means**.

> **Rationality is no longer a credential.**
> **It is a capacity that must be continuously developed by everyone.**

Within the logic of the **Human Growth Center**:
- Rationality is not a filter
- It is a growth instrument
- Not a symbol of exclusion
- But a shared public resource

IX. Conclusion: When Rationality No Longer Needs Gatekeepers

French education represents one of the most self-aware rational projects in modern civilization.

But in the age of AI:

When thinking can be amplified, distributed, and collaboratively developed,
when the state is no longer the sole source of intellectual legitimacy,

any system that depends on:
- Extreme selection
- Symbolic scarcity
- Elite-exclusive pathways

will inevitably destabilize.

Not because France was wrong—

but because:

Rationality no longer needs exclusion to prove its value.

The AI era does not require
fewer people thinking better.

It requires:

More people continuously growing into rational, responsible, and autonomous human beings.

Chapter 6 | Chinese Education: The Human Cost of Extreme Efficiency

What Is Education, Really?

What is education?

It is a word repeatedly used throughout human history—yet rarely questioned in any serious way.

I come from China's education system. The **Gaokao** (the national college entrance examination) is often described by observers inside and outside China as a "universal" and "most fair" institution, as if it offers everyone an equal path toward upward mobility. Yet very few people ask the deeper questions:

Where did it come from? Who was it designed for? And where does it ultimately lead human beings?

In truth, the Gaokao was not built out of concern for "human development." It is an institutional design created fundamentally to serve **political order and social stability**.

Through the logic of score worship, it locks nearly all adolescents inside classrooms for years, confining what should be society's most powerful force for renewal and transformation inside a system-cage called **test preparation**.

The core goal of this education system has never been to liberate individuals. It has been to preserve the rigidity of the existing political and social structure—ensuring that vested interests can be extended across generations.

More deceptively, it wraps itself in the language of "equal opportunity," manufacturing the illusion of "Gaokao fairness" within a system where political rights and social rights are already profoundly unequal. But in a society that is structurally unequal at its foundations, **where can true fairness come from?**

64

The consequence of score worship is that only a small minority of students "leap over the dragon gate." And these selected "elites" are often assigned an even more hidden function:

They become **living proof of the system's legitimacy.**

Their success stories are repeated again and again, used to persuade countless "losers" to keep believing in the justice of scores—rather than asking the more fundamental question:

> True equality is not equality of exam scores.
> True equality is equality of political rights and social rights for every person.

> The irony is that those who succeed in the leap often quickly forget where they came from and instead become defenders of the very system that crushed them.

> Because they have become beneficiaries—and they even learn to manage, filter, and control others using methods resembling the old power itself, turning into a new kind of "people above people."

This is the essence of Chinese education.

I. The Fundamental Logic of Chinese Education: Unified Exams as Social Ranking and Political Stabilization

If the underlying logic of Chinese education can be summarized in one sentence, it is this:

> Through nationally unified, highly standardized examinations, the population is ranked on a massive scale at low cost and with high predictability—thus maintaining political and social stability.

> This is not simply an "education problem."

It is a governance problem.

In a society where:
- Political participation is highly restricted
- Rights cannot compete freely
- Resource distribution is highly centralized

there must exist a **non-political mechanism of ranking** that makes people "accept their position."

The Gaokao is precisely such a mechanism.

It transforms energy that could otherwise become:
- Political demands
- Institutional criticism
- Social confrontation

into:
- Score competition
- Individualized internal war (involution)
- Self-blame

Thus completing a highly successful project of **de-politicizing society**.

II. Why Was Chinese Education Historically "Effective"?

We must acknowledge: in certain historical stages, Chinese education did produce real "success."

1 Rapid Mass Production of Basic Skills

For decades after Reform and Opening Up, China's main tasks were:
- Industrialization
- Urbanization
- Labor transfer on a huge scale

Chinese education was extremely effective at one thing:

> Training hundreds of millions of people—at the lowest cost and at the fastest speed—to acquire basic literacy, numeracy, and rule-following capacity.
>
> This provided critical human foundations for:
> * Manufacturing expansion
> * Infrastructure acceleration
> * Urban operations

2 A Narrow but Real Pathway from Rural to Urban Mobility

The Gaokao did provide a narrow but real upward channel for some rural and low-income families.

These success cases were constantly magnified, retold, and mythologized—becoming the system's most important moral shield:

> "Look, the system is fair. You just didn't work hard enough."

> But the problem is obvious:

> The existence of a few upward channels does not prove the justice of the overall structure.

III. The Real Function of Chinese Education: Not Cultivating Humans, but Pressing Humans Down

When we strip away moral narrative and emotional packaging, the true function becomes visible.

1 Humans Are Systematically Reduced into "Calculable Objects"

In Chinese education:
- Human ≈ score
- Value ≈ ranking
- Potential ≈ predictable output

Everything non-quantifiable becomes:
- Useless
- Dangerous
- A disruptive factor

Including:
- Emotion
- Value judgment
- Independent thinking
- Questioning authority

2 Creativity Is Treated as an "Instability Factor"

Creativity implies:
- Not following standard paths
- Challenging fixed answers
- Redefining problems

But Chinese education pursues the opposite:
- Standard answers
- Unified pace
- Controlled outcomes

Therefore creativity is not merely ignored—it is actively suppressed.

3 Dignity Is Systematically Stripped Away

Long-term test training doesn't just create "pressure."

It gradually shapes a personality structure:
- Becoming addicted to evaluation
- Terrified of mistakes
- Outsourcing self-worth to authority
- Developing deep shame around failure

This is not a side effect.

It is a necessary product of education operating under a governance logic.

IV. The Time Scam: "Happiness Postponed to Forever"

The cruelest part of Chinese education is not suffering itself, but how it manipulates time.

Children are told repeatedly:
- Suffer now, life will be better later
- After Gaokao you will relax
- After university you will be free
- After work you will be happy

But the reality is:
- After Gaokao comes graduate exams
- After graduate exams comes employment
- After employment comes mortgage
- After mortgage comes anxiety about aging

Happiness is delayed again and again—until life ends.

This is a time-based technology of control.

V. The Hidden Massive Cost: The Destruction of Social Potential

The cost is not carried only by "failures."

1 Late Bloomers Are Eliminated Systematically

Many people:
- Understand themselves only after 20
- Find direction only after 30
- Create real value through cross-domain life paths

But Chinese education completes at around 18:
- Identity labeling
- Social positioning
- Psychological fixing

Late awakening is not treated as a possibility—it becomes a "failure label."

2 Society's Innovation Capacity Is Flattened

A society that depends long-term on:
- Memorization
- Drill practice
- Standard answers

is unlikely to produce:
- Original innovation
- Deep value creation
- Civilizational breakthroughs

This is why China can become extremely powerful in scale—yet remain structurally limited in originality and global intellectual output.

VI. The Fundamental AI-Era Break: Chinese Education Becomes Fully Invalid

AI is not merely a challenge to Chinese education.

It is a fatal negation.

1 Memorization and Drill Value Collapse to Zero

AI comprehensively outperforms humans in:
- Memory
- Calculation
- Pattern recognition
- Standard problem-solving

Chinese education's "core competitive strength" becomes meaningless labor almost overnight.

2 The Authority of Standard Answers Collapses

When anyone can:
- Obtain answers instantly
- Understand from multiple angles
- Integrate across disciplines quickly

the concept of "the one correct answer" collapses at the technical level.

3 Obedience-Type Personality Becomes a Liability

What the AI era truly requires is:
- Judgment
- Direction
- Value-based choice
- Self-driven learning

These are precisely what Chinese education has suppressed for decades.

So your diagnosis is accurate:

In the AI era, Chinese education is like an ox cart in the age of airplanes.
Not just slow—its direction is wrong.

VII. Why "Reforming the Gaokao" Is Meaningless

Under the current framework:
- Reducing burden
- "Quality education" slogans
- Digital education upgrades

are rhetorical decorations.

The Gaokao must be completely abolished.

As long as:
- The rights structure remains unchanged
- Education continues to function as a social ranking machine
- Humans remain treated as governance objects

Chinese education cannot truly transform.

VIII. This Is Not an Education Issue, but a Civilizational Choice

The problem is not:
- Teachers not working hard
- Parents being irrational
- Students lacking intelligence

The problem is:

From the beginning, it was not designed for "a person becoming themselves."

In the AI era, this model has reached its historical end.

The true dividing line is no longer:
- Who scores higher
- Who can endure more suffering

But:

Who can abandon defining humans by exams—
and instead clear the path for human unfolding.

This transformation will decide not only the fate of education,
but whether Chinese society can enter a truly modern civilizational stage.

IX. Why Are the "Successful" Often the Most Loyal Defenders? The Psychology of the Selected

A crucial but long-ignored phenomenon:

The most passionate defenders of China's education system are often those who "succeeded" within it.

This is not accidental. It is psychologically predictable.

1 Sunk Cost: Denying the System = Denying the Self

Success in Chinese education often means:
- More than a decade of extreme obedience
- Long-term suppression of interest, emotion, and self
- Betting one's entire youth on a single path

If they admit "the system itself is wrong," they are admitting:

> "I spent the most precious years of my life inside a wrong machine."

> Most people cannot bear that.

> So they choose the safer route:
- Rationalize suffering
- Sanctify success
- Justify the system

2 Survivor Bias Disguised as Moral Truth

China depends heavily on a narrative:

> "If you work hard enough, you can succeed too."

> But this is survivor bias moralized.

> Successful people:
- Cannot see the silent massive population eliminated
- Assume their path is universal
- Translate luck and structural advantage into "personal virtue"

Eventually they say the most harmful sentence:

> "I could do it—why can't you?"

3 Authorized Second-Stage Oppression

Once they become beneficiaries, they are given roles:
- Teachers
- Managers

- Evaluators
- Employers

They begin to:
- Filter by scores
- Judge by credentials
- Measure by obedience

And unconsciously reproduce the oppression they once endured.

This is not moral corruption—it is structural reproduction.

X. Parents–Schools–State: A Stable Three-Party Co-Production System

Chinese education is stubborn because it is not simply imposed by one power.

It is a system where three parties reinforce each other.

1 The State: Replacing Rights Distribution with Exams

For the state, the Gaokao's greatest value is:
- Turning class struggle into "technical competition"
- Converting dissatisfaction into individual failure
- Preventing rights questions from entering public discussion

Exams become a governance tool of de-politicization.

2 Schools: Executing Ranking, Not Cultivating Humans

In this structure, schools do not truly exist to educate.

They exist to:
- Train test-taking ability
- Maintain ranking order
- Output predictable results

Teachers' ideals are irrelevant, because the system rewards only one KPI:

Admission rates.

3 Parents: Fear-Driven Collaborators

Parents often do not believe the system is good.
But they fear:
- Their child "falling behind"
- Being eliminated by society
- Becoming "failed parents"

So they speak the tragic line:

"I know it's bad—but there's no alternative."

Fear turns them into some of the system's most loyal executors.

The Three-Party Loop Produces:
- Stability for the state
- Evaluation safety for schools
- Psychological comfort for parents

And the only sacrifice is:

The child's full unfolding as a human being.

XI. Long-Term Consequences: A Manufactured Personality Structure

Chinese education shapes not only skills—but a full personality model.

1 External-Validation Dependence

Long-term exam culture creates people who:
- Cannot judge value internally
- Need external approval to act
- Fear rejection deeply

This leads to:
- Fear of entrepreneurship
- Expression anxiety
- Avoidance of innovation risk

2 Reflex Obedience to Authority

From childhood, students are trained:
- Standard answers exist
- Authority is always correct
- Questioning equals danger

This clashes directly with AI-era requirements:
- Defining problems
- Handling conflicting judgments
- Making value-based choices

3 Emotional Numbness Toward Happiness

When people are trained:
- Endure first
- Wait later
- "Talk about happiness in the future"

Eventually they lose the ability to sense happiness.

Not because they are not unhappy—
but because they no longer know what happiness is.

XII. A Fundamental Contrast with the Nordic Model

If China's core is:

"Flatten first, then filter."

Nordic education is the opposite:

"Protect unfolding first, let differences emerge naturally."

The real difference is not simply whether there are exams.

It is:
- Whether late awakening is allowed
- Whether nonlinear growth is respected
- Whether humans are treated as ends rather than tools

This explains why:

China excels at "copying successful models,"
but struggles to produce original civilizational innovation.

XIII. AI Era: A Civilizational Mismatch

The AI era amplifies not efficiency—but difference:
- AI amplifies those with judgment
- AI eliminates those who only execute
- AI rewards those who can continuously rebuild themselves

But Chinese education systematically:
- Erases difference
- Suppresses judgment
- Punishes deviation

This is not "one step behind."

It is the opposite direction.

XIV. Final Conclusion: An Unavoidable Civilizational Choice

Chinese education can no longer be solved by:
- Tweaking exams
- Adding AI
- Changing textbooks

Because its essence is:

An education project designed to produce controllable populations—
not a growth system designed to produce whole human beings.

Continuing this model in the AI era will only produce:
- Massive waste of potential
- Collapse of social creativity
- Widespread psychological depression

The real choice is only one:

Continue using education to manage humans—
or begin clearing the path for human growth.

Chapter 7 | Indian Education: Education as the Only Lottery to Escape Fate

What Happens When Education Is Asked to Do the Impossible

Education, in theory, is meant to cultivate human potential.

In India, it has been burdened with a far heavier and more tragic role:

education has been turned into the only socially legitimate escape route from fate.

Not one path among many.
Not a developmental system.
But a **lottery**.

A lottery asked to overcome:
- millennia of caste stratification
- religious fragmentation
- colonial language hierarchy
- extreme economic inequality
- and weak state capacity

No education system can survive such a burden intact.

I. The Core Logic: Extreme Competition as a Substitute for Social Reform

The foundational logic of Indian education can be summarized in one sentence:

> When society cannot guarantee dignity, security, or mobility, education is forced to function as a zero-sum escape mechanism.

Unlike China, where exams are used to **stabilize society**,
India uses exams to **filter hope**.

The system does not promise fairness.
It promises **a miracle**.

Education as a Fate-Breaking Device

For millions of Indian families, education is not about:
- self-discovery
- intellectual growth
- civic development

It is about:
- escaping caste
- escaping village immobility
- escaping religious marginalization
- escaping generational poverty

Education is not preparation for life.

It is a **desperate wager against destiny**.

II. Why Extreme Competition Became Inevitable

India did not choose extreme educational competition by ideology.

It was cornered into it by structure.

1 Caste: A Social Hierarchy That Education Is Asked to Undo

The caste system—formal or informal—still determines:
- social status

- marriage opportunities
- occupational ceilings
- local power dynamics

Education became the **only socially acceptable way** to claim legitimacy beyond caste.

But when education is asked to compensate for structural injustice, it becomes brutal.

Everyone is told:

> "If you fail, it is because you did not study hard enough."

> Structural oppression is converted into **individual failure**.

2 Religion and Fragmented Identity

India is not one social body.
It is a **constellation of communities**.

Religion, language, region, and identity shape access to:
- schools
- networks
- safety
- opportunity

Education is imagined as a neutral ladder—but in reality, it is climbed from profoundly unequal starting points.

This produces:
- resentment
- anxiety
- politicization of education itself

3 Colonial Legacy: English as the Gatekeeper

One of the most decisive fractures in Indian education is **language**.

English proficiency often determines:
- access to elite institutions
- access to global labor markets
- access to knowledge capital

English is not just a language.
It is a **class divider**.

Those without early exposure are effectively excluded—regardless of intelligence.

III. The Brutal Mechanics of Indian Educational Selection

India's educational filtering operates through:
- IITs
- IIMs
- medical entrance exams
- civil service examinations

Each functions as a **massive elimination funnel**.

Extreme Competition at Unprecedented Scale
- Millions compete
- Thousands survive
- Hundreds thrive

Success rates often fall below **1%**.

This is not selection.
This is **mass extinction of aspiration**.

The Psychological Cost of Constant Near-Failure

Unlike systems with early tracking, India keeps hope alive—until it crushes it.

Many students spend:
- years in coaching centers
- entire family savings
- repeated exam attempts

Only to fail **again and again**, often by microscopic margins.

Failure here is not educational.

It is existential.

IV. What the System Actually Produces (And What It Sacrifices)

The Real Successes

India undeniably produces:
- world-class engineers
- exceptional mathematicians
- elite technologists
- globally competitive professionals

These individuals:
- power Silicon Valley
- staff global research labs
- anchor multinational corporations

They are real.
They are remarkable.
They are also **statistical anomalies**.

The Hidden Cost: Structural Mass Failure

For every global success story, there are:
- millions labeled "not good enough"
- millions whose self-worth collapses
- millions whose families are financially ruined
- millions whose anxiety becomes chronic

Education becomes a **societal anxiety amplifier**.

Hope is rationed.
Failure is normalized.
Mental health is collateral damage.

V. Education as a Social Pressure Cooker

Indian education does not merely educate.

It **concentrates pressure**.
- parental expectations
- community honor
- religious identity
- economic desperation

All converge onto a single exam score.

This produces:
- extreme stress
- student suicides
- widespread burnout
- loss of intrinsic curiosity

The system survives not because it works—but because there is **nothing else**.

VI. Why This Model Cannot Survive the AI Era

The Indian educational lottery depends on one assumption:

> A tiny elite controls scarce technical knowledge and high-paying opportunities.

> AI destroys this assumption at its root.

1 Knowledge Is No Longer Gatekept

AI provides:
- advanced explanations
- programming assistance
- problem-solving support
- research synthesis

Access no longer depends on elite institutions.

2 Tools Are No Longer Reserved for the Few

Capabilities once requiring:
- top universities
- expensive labs
- elite mentorship

are increasingly available to individuals and small teams.

3 Extreme Filtering Loses Economic Justification

When:
- skill acquisition accelerates
- talent can emerge late
- creativity outweighs credentials

a system designed to reward **only 0.1%** becomes economically irrational.

VII. The Collapse of the Lottery Illusion

In the AI era:
- success is less linear
- careers are reconfigurable
- value is continuously rebuilt

A lottery-based education system:
- cannot accommodate late bloomers
- cannot absorb re-skillers
- cannot support distributed innovation

It produces despair—not adaptability.

VIII. The Deeper Truth: Education Cannot Fix Civilizational Fault Lines

Indian education has been asked to fix:
- caste injustice
- religious division
- language inequality
- economic stratification

No education system can compensate for all of these.

When education is forced to do so, it becomes cruel.

IX. Conclusion: When Education Becomes a Lottery, Humanity Loses

The tragedy of Indian education is not that it produces elites.

It is that it **sacrifices humanity** to produce them.

In the AI era, the question is no longer:
- "Who can survive the competition?"

But:
- "How can every human continuously grow?"

A system built on extreme filtering, extreme pressure, and extreme inequality **cannot answer this question**.

The lottery must end.

Not because India lacks talent—
but because it has too much human potential to waste.

Chapter 8 · Latin America: When Human Ideals Exist, but Institutions Cannot Carry Them

What Happens When a Society Truly Respects the Human Being—

Yet Lacks the Capacity to Sustain Human Destiny?

Among all global education regions, Latin America occupies a uniquely misunderstood position.

It is not like China, with its highly centralized, ultra-efficient governance machine.
It is not like India, where education has been reduced to a brutal lottery to escape fate.
Nor is it like Europe, where institutional rationality has accumulated over centuries.

Instead, Latin America reveals a paradoxical but internally consistent condition:

> **Human-centered ideals genuinely exist,**
> **but state and institutional capacity cannot sustain them.**

> This is a civilization where **moral intention precedes structural ability**.

I. Core Logic: Human Values Are Real, State Capacity Is Not

If one sentence were to capture the educational logic of Latin America, it would be this:

> **Society does not deny human dignity or difference—**

89

but the state lacks the power to translate respect into stable outcomes.

This distinguishes Latin America fundamentally from China and India.

Here:
- The problem is not systematic oppression
- The problem is systemic fragility

Education is not designed to discipline the population— it is simply unable to hold it.

II. Education as Powerless Goodwill

In most Latin American countries, education is not constructed as:
- An extreme sorting machine
- A hyper-competitive battlefield
- A total obedience apparatus

On the contrary, at the level of declared values:
- Equality is emphasized
- Public education is defended
- Emotional development is acknowledged
- Competition is often morally questioned

Yet a fatal gap persists:

Values exist without enforcement capacity.

III. Why Education Fails to Change Destiny

1 Weak State Capacity: Not Malice, but Inability

The fundamental limitation of Latin American education is not ideology—it is capacity.

Common features include:
- Chronically unstable education funding
- Undertrained and underpaid teachers
- Severe infrastructure inequality
- Educational policy reversal with every political cycle
- Corruption eroding implementation

As a result:

> Education is not oppressive—
> it is structurally unsustainable.

2 Extreme Inequality Neutralizes Education's Promise

Latin America is among the most unequal regions on Earth.
- Vast gaps between urban and rural areas
- Deep separation between elites and informal populations
- Large informal economies
- Persistent violence, gangs, and drug-linked power structures

Education is therefore burdened with an impossible expectation:

"Study hard, and you can escape your fate."

But reality contradicts this promise:
- Labor markets cannot absorb graduates
- Legal systems cannot protect upward mobility
- The middle class remains fragile and reversible

Education loses its credibility as a path.

3 The Latin Cultural Paradox: Human Warmth, Institutional Fragility

Latin American societies strongly emphasize:
- Emotional expression
- Human connection
- Family and community
- Social empathy

This preserves immense cultural vitality.

But it also generates a long-term weakness:

Strong emotional culture, weak institutional culture.

In education, this appears as:
- Empathy for children
- Reluctance toward extreme pressure
- Resistance to brutal elimination

Yet simultaneously:
- Weak evaluation systems
- Short policy horizons
- Inconsistent standards
- Limited accountability

IV. Education in Suspension: Neither Success Nor Collapse

Latin American education does not collapse outright.

It **hovers**.

It promises, but cannot deliver.

1 Education Does Not Reliably Alter Life Trajectories

For millions:
- School ≠ stable employment
- Diploma ≠ social security
- Degree ≠ upward mobility

A collective psychological adaptation emerges:

"Education is good—but do not depend on it."

This is a dangerous civilizational signal.

2 Chronic Disconnection Between Hope and Reality

Political discourse remains idealistic.
Schools teach progress narratives.
Reform language persists.

Yet social reality offers:
- Few dignified jobs
- Weak absorption of youth
- Insecure futures

Hope and reality drift apart.

V. The Underestimated Strengths of Latin America

Despite institutional failure, Latin America is not a failed civilization.

It preserves capacities that other systems have already destroyed.

1 Creativity Beyond Institutions

Latin America exhibits extraordinary strength in:
- Music
- Literature

- Visual arts
- Sports
- Social movements

This creativity is often:
- Non-elite
- Non-credentialed
- Rooted in lived experience

It is **undomesticated creativity**.

2 Emotional Intelligence and Empathy

Latin cultures maintain:
- High emotional literacy
- Strong interpersonal bonds
- Community-based resilience

These capacities are rare in highly optimized industrial systems—
and profoundly valuable in the AI age.

3 Social Resilience Under Instability

Decades of crisis have produced a distinctive skill:

The ability to live without guarantees.

This is not motivational rhetoric—
it is adaptive intelligence.

VI. The Systemic Cost of Unfulfilled Promises

1 Youth Disillusionment and Cynicism

When education repeatedly fails to deliver, young people learn:

"Do not take institutions seriously."

Consequences include:
- Declining civic participation
- Distrust of authority
- Short-term survival thinking

2 Informal Pathways Replace Educational Ones

In many regions:
- Informal economies
- Migration
- Gangs
- Grey markets

Become more rational than schooling.

This is not cultural failure—
it is structural necessity.

VII. The AI Breakpoint: A Rare Historical Opening

Unlike China or India, Latin America may encounter **AI as an opportunity rather than a threat**.

1 AI Can Bypass Weak Institutions

AI enables:
- Direct access to knowledge
- Global collaboration
- Skill acquisition without credentials
- Learning without state mediation

For the first time in history:

Human growth may no longer depend on national capacity.

2 Creativity × AI: A Structural Match

Latin America already possesses:
- Narrative ability
- Expressive intelligence
- Cultural imagination

AI lowers technical barriers and amplifies output.

This combination is structurally powerful.

3 Decentralized Growth Fits Local Reality

AI-based learning:
- Does not require stable schools
- Does not depend on certification
- Does not assume formal labor markets

It aligns with existing social conditions.

VIII. The Real Risk: Missing the Window Again

Opportunity does not guarantee outcome.

Risks include:
- AI monopolized by capital
- Education discourse remaining obsolete
- States failing to provide minimal digital access

If so, history may repeat itself.

IX. Conclusion: Latin America Is Not a Failure—It Is Delayed by History

Latin America's core problem has never been contempt for human beings.

It is:
- Respect without infrastructure
- Recognition without continuity
- Aspiration without execution

In the AI era, a rare possibility emerges:

Human growth without perfect institutions.

If AI is treated as a public capability rather than an elite asset,
Latin America may become:

One of the first real laboratories of post-education human development.

Chapter summary :

- **China**: Suppressive efficiency
- **India**: Lottery-based escape
- **Latin America**: Benevolent incapacity

Together they reveal a shared truth:

When education is forced to repair society, it inevitably fails.

AI finally allows humanity to ask the real question:

What if society stopped asking education to fix inequality—

and instead reorganized itself to allow human growth?

Chapter 9 · Finland and Nordic Education: The Most Successful Pre-AI Model—But Not the End Point

When Education Truly Respected the Human Being—

What Did It Prove, and What Did It Still Fail to Answer?

Before the arrival of artificial intelligence, humanity conducted many large-scale experiments in education.
Most of them failed—often brutally.

China maximized efficiency at the cost of human dignity.
India turned education into a survival lottery.
Latin America preserved human warmth but lacked institutional capacity.

Against this global backdrop, **Finland and the Nordic countries stand apart**.

They represent the **most humane, stable, and ethically coherent education system humanity achieved before AI**.

Not perfect.
Not universal.
But historically extraordinary.

I. Core Judgment: The Closest Pre-AI Approach to Respecting the Human Being

If one had to name a single region where education came closest to respecting human life *as it is*, rather than reshaping it into a social instrument, it would be the Nordic world.

Nordic education demonstrated something radical by historical standards:

Human dignity does not have to be sacrificed for social order.

This alone makes it a civilizational milestone.

II. What Nordic Education Proved—Once and for All

1 Low Control Does Not Produce Chaos

Contrary to centuries of educational fear, Nordic systems showed that:
- Fewer standardized exams
- Minimal ranking
- High teacher autonomy
- Strong student trust

did not result in social breakdown.

Instead, they produced:
- Stable societies
- High literacy
- High civic participation
- Low crime
- High trust

This disproved a foundational assumption of industrial education:

That human beings must be tightly controlled in order to function.

2 Respecting Difference Does Not Mean Lowering Quality

Nordic education rejected:
- Early tracking
- Harsh selection
- Permanent labeling

Students were allowed to:
- Develop at different speeds
- Fail without permanent punishment
- Discover abilities late

Yet outcomes remained strong.

This destroyed another long-held myth:

> **That equality of dignity inevitably leads to mediocrity.**

Nordic systems showed that:
- Supporting the weakest does not destroy excellence
- Human variability is not an obstacle—it is a resource

3 Happiness Does Not Undermine Society

Perhaps the most radical finding of all:

> **Human happiness is not incompatible with social stability or productivity.**

Nordic countries demonstrated that:
- Less anxiety did not reduce competence
- Fewer exams did not destroy discipline
- Emotional well-being did not weaken responsibility

This was historically unprecedented.

For the first time, education proved that **life could be livable while remaining functional**.

III. Why Nordic Education Worked (When Others Didn't)

Its success was not accidental.

1 High Institutional Trust

Nordic societies are characterized by:
- Low corruption
- Strong rule of law
- Stable public institutions

Education was not required to:
- Fix society
- Replace justice
- Compensate for political failure

This removed immense pressure from schools.

2 Strong Welfare Foundations

Because healthcare, housing, and social security were guaranteed:
- Education was not a life-or-death gamble
- Failure did not equal social annihilation

This allowed education to be **developmental rather than defensive**.

3 Teachers as Trusted Professionals

Nordic teachers are:
- Highly trained
- Well-respected
- Given autonomy

They are not micromanaged by:
- Test metrics

- Ranking pressure
- Political panic

This enabled genuine pedagogy rather than mechanical instruction.

IV. The Structural Limit: Why Nordic Education Is Still Pre-AI

Despite all its achievements, Nordic education remains **historically bounded**.

It belongs to the **best possible version of the old paradigm**.

1 Schools Still Function as Central Learning Institutions

Even in Nordic systems:
- Learning is still organized around schools
- Time is still segmented
- Childhood is still institutionally managed

Education remains a **separate phase of life**, not fully integrated into society.

2 Teachers Are Still the Primary Cognitive Guides

Though more humane, the structure remains:

Teacher → Student

Even when supportive, this assumes:
- Knowledge scarcity
- Human-limited attention
- Fixed instructional bandwidth

These assumptions collapse under AI.

3 Learning Still Occurs "Before Life"

Nordic education still assumes:
- Learn first
- Live later

Even if the process is kinder, the sequence remains intact.

AI disrupts this completely.

V. The AI Breakpoint: When the Nordic Model Reaches Its Limit

AI introduces a rupture that even Nordic education cannot absorb without transformation.

1 Knowledge Is No Longer Scarce

The Nordic model still relies on:
- Curriculum planning
- Teacher knowledge mediation

AI removes this constraint entirely.

2 Personalization Exceeds Human Capacity

Even the best teacher cannot:
- Track thousands of micro-learning trajectories
- Adjust in real time
- Scale empathy infinitely

AI can.

3 Learning Escapes Institutional Time

AI enables:
- Lifelong learning
- Contextual learning
- Learning embedded in real activity

This renders the "school phase" historically provisional.

VI. What Nordic Education Teaches the Future—And What It Cannot Provide

What It Teaches Us

Nordic education proves that:
- Humans do not need to be broken to function
- Fear is not a prerequisite for learning
- Dignity can coexist with competence

These lessons are permanent.

What It Cannot Do

It cannot answer the AI-era question:

What happens when learning no longer needs institutions at all?

VII. From Nordic Education to the Human Growth Paradigm

Nordic systems demonstrated:

Less control can work.

AI now forces the next question:

**Can we remove control entirely—
without losing coherence?**

This leads beyond education itself.

VIII. Conclusion: Nordic Education Was the End of One Era, Not the Beginning of the Next

Nordic education represents:
- The ethical peak of industrial education
- The most humane pre-AI model ever built

But it is still:
- School-based
- Teacher-centered
- Phase-bound

AI opens something fundamentally different:

A world where learning is continuous, distributed, and inseparable from life itself.

The Nordic model answers the question:

"How human can education be?"

The AI era asks a new one:

"Do we still need education at all?"

That question defines the next chapter of human civilization.

Chapter 10 The Ultimate Comparative Chapter | Africa × Nordic

The Beginning and the Upper Limit of Pre-AI Education

Part I

Africa — Inherited Schools and Interrupted Human Potential

1. A Fact That Must Be Stated First

Africa's education systems did **not** grow organically from African civilizations.

Unlike Europe, East Asia, or parts of the Middle East, the dominant school structures across most African countries did not emerge from internal reflection on how humans should grow or learn. Instead, they were:

- forcibly transplanted during the colonial era
- designed to serve administration, religious conversion, and basic governance
- aimed at producing *manageable people*, not *developing human potential*

From the very beginning, the core problem of African education was therefore **not performance**, but **misalignment**.

This system was **never designed for African societies** in the first place.

2. Core Institutional Logic: External Recognition Over Local Reality

The implicit objectives of mainstream African education systems can be summarized as three priorities:

1. **External legitimacy**
Curricula, exams, and diplomas must resemble European or "international" systems.

2. **Elite exit channels**
Education functions as a pathway out of local society and into the global system.

3. **Minimal social order**
Basic literacy, discipline, and administrative usability.

What these systems largely ignore are three essential questions:

- Does schooling address real local problems?
- Is learning connected to community survival and daily life?
- Are individual potentials recognized and allowed to grow?

3. What African Education Did Achieve (This Must Be Acknowledged)

To deny all achievements would weaken the analysis. African education systems did succeed in several limited ways:

✓ Basic literacy and social stability
In some regions, schools raised literacy rates and provided a basic framework of order.

✓ Elite professional pathways
Africa produced doctors, engineers, scholars, and international officials who function effectively in global systems.

✓ Minimal nation-building tools
For newly formed states, schools symbolized institutional presence.

But all these achievements share one feature: they serve a small minority and **cannot reproduce themselves at scale**.

4. The Systemic Cost: What Was Sacrificed Was Not Skill, but Life Itself

4.1 Total Disconnection from Real Life
Across large parts of Africa:
- agriculture, craftsmanship, community governance, and local ecology
are treated as "non-educational"
- Western curricula are positioned as the *only legitimate path*

As a result, what children learn in school often has **almost no overlap** with the lives they must actually live after leaving it.

4.2 Education as a Mechanism of Elite Extraction
"Successful education" frequently means:
- leaving the village
- leaving the community
- ultimately leaving the country

Education becomes a system that **drains local societies of their most capable people**.

4.3 Systemic Blindness to Africa's Real Strengths
African societies have long cultivated:
- strong rhythmic intelligence
- improvisational creativity
- visual and embodied cognition
- oral knowledge transmission
- intergenerational learning

Within formal school systems, these capacities are labeled:

"non-academic"
"informal"
"irrelevant"

This is not a lack of ability —
it is **institutional blindness**.

5. Africa's AI Breakpoint: Not Lagging Behind, But Leaping Forward

This is the chapter's most critical judgment.

Unlike many regions, Africa possesses four unique conditions:
1. relatively weakly entrenched education bureaucracies
2. extremely young populations
3. widespread informal learning cultures

4.	fewer illusions about the old education model

This means:

Africa may not be the last to enter the AI era
—
it may be the **first capable of skipping the full education stage altogether**
and moving directly into a **post-education civilization**.

6. Africa's Natural Alignment with Human Growth

Across many African communities, unacknowledged learning models already exist:
- learning across generations
- observation → imitation → practice
- responsibility before theory
- survival as learning

These align perfectly with the core principle:

Humans are not educated — they grow through life itself.

AI does not make Africa "more Western."
For the first time, it allows Africa to **stop denying its own civilizational experience**
and translate it into a future-oriented structure.

7. Africa's Civilizational Position

Africa's education challenge was never "lack of effort."
It was being forced to inherit a school structure
fundamentally incompatible with its social reality.

8. Africa's Unique Position in the Global Education Map
- China represents **the extreme efficiency endpoint** of education
- Nordic countries represent **the human-centered upper limit of pre-AI education**
- Africa represents something else entirely:

> a region **least shaped by successful schooling,** and therefore **most capable of entering the post-education era first**

> Where others suffer because education worked *too well,*
> Africa's system never fully merged with life.

> Paradoxically, this becomes an advantage once AI reshapes the meaning of knowledge and learning.

> Africa may become the **first continent that no longer needs education,**
> only **conditions for human growth.**

Part II — Africa × Nordic Countries

Why Africa and the Nordics Must Be Placed on the Same Civilizational Map

In global education discourse,
Africa is often framed as a **"lagging sample,"**
while the Nordic countries are praised as an **"ideal model."**

In the age of AI, this ranking itself is fundamentally wrong.

Africa and the Nordic countries do **not** stand on a scale of advancement.
They stand on the **same historical axis,**
but at **two opposite civilizational positions**:
    •   **Africa**
    → A human region **never fully shaped by modern education systems**
    •   **Nordic countries**
    → The **most successfully realized form of human-centered education in the Pre-AI era**

Placing them side by side is not about comparison or judgment.
It is about answering a much deeper question:

As "education" approaches the end of its historical role,
where did human civilization come from—
and where is it going next?

Section I — Africa: A Continent Education Never Fully Captured

Africa — Where Education Never Fully Took Hold

1. Africa Is Not an "Education Failure," but a Region Never Fully Educated

Unlike East Asia or Europe,

most African societies never entered a phase where **school systems fully determined life trajectories**.

Across vast regions of Africa:
 • Learning has always occurred largely outside schools
 • Capability is formed through real-life practice
 • Life paths are not completely locked by exams or credentials

This is neither an advantage nor a disadvantage.
It is a **rare historical condition**.

2. Schools as External Structures, Not Civilizational Organs

In most African countries:
 • Schools are **not** the core of community life
 • Learning is **not** the central axis of identity
 • Diplomas are **not** the primary source of dignity

Schools function mainly as:
 • Administrative extensions
 • Interfaces with international systems
 • Exit channels for a small elite

As a result, education never deeply redefined
what a "successful life" means,
unlike in Europe or East Asia.

114

3. Africa's Overlooked Human Capability Structure

African societies have long preserved capabilities that
modern education suppressed,
yet AI has made newly essential:
- Contextual judgment
- Improvisational creativity
- Intergenerational collaboration
- Embodied and sensory intelligence
- Community responsibility over abstract
rules

Traditional education labeled these as:

"Non-academic," "informal," or "irrelevant."

In the AI era, they are precisely the abilities
least automatable and most human.

This is not a lack of capability—
it is **institutional blindness**.

4. Africa's Real Potential: Direct Entry into the Post-Education Era

Because Africa was never fully "education-shaped,"
it possesses a rare civilizational advantage:

**It does not need to dismantle a massive education
bureaucracy
to move forward.**

With AI support:
- Knowledge barriers collapse rapidly
- Learning no longer depends on schools

115

- Communities re-emerge as growth units

Africa may not be a "catch-up continent,"
but a **civilizational leapfrogging continent**.

Section II — The Nordics (Finland): The Upper Limit of Pre-AI Education

Nordic Education — The Highest Point Before the Break

5. What Nordic Education Truly Achieved

Nordic countries—especially Finland—accomplished something historically rare:
- ☑ Low testing, low coercion, yet stable societies
- ☑ Respect for difference without educational collapse
- ☑ High well-being coexisting with social functionality

This represents the **highest achievement of human-centered education before AI**.

6. Nordic Success Also Marks Education's Structural Ceiling

Yet clarity is required:

The Nordic system is still an **education civilization**,
not a post-education one.

It retains assumptions that will inevitably fail in the AI era:

116

- Schools remain central institutions
- Teachers remain structural authorities
- Learning is still framed as "before real life"
- Life paths are still implicitly guided

These are not flaws—
they are **limits of a historical stage.**

7. What the Nordics Proved—and What They Cannot Answer

The Nordics proved this to the world:

Humans can be educated well
without fear, competition, or exam violence.

But the AI era poses a deeper question:

**When knowledge no longer needs to be taught,
is "education" itself still necessary?**

This is a question Nordic systems cannot answer—
not because they failed,
but because the question did not exist when they were
built.

Section III — Civilizational Contrast: Origin ×
Upper Limit

Dimension	Africa	Nordics
Education penetration	Very low	Very high
School control over life	Weak	Moderate but systemic

Life-path lock-in	Minimal	Minimal (but present)
Source of well-being	Community & relations	Institutional security
AI impact	Leapfrog restructuring	Structural transformation

Section IV — The Answer Lies Beyond Both

Africa is not the answer.
The Nordics are not the answer.
- Africa lacks institutional safeguards
- The Nordics remain Pre-AI in structure

Yet together, they reveal a single truth:

**Human beings do not need to be heavily engineered
to become whole.**

If Nordic education proved that
education does not need to oppress,
Africa reminds us that
human life never depended entirely on education to begin with.

AI is the force that finally allows these two historical truths to converge.

Conclusion — From Education to Human Growth Center

Placing Africa and the Nordics on the same civilizational map reveals what was long invisible:

> **Education is neither the origin nor the destiny of human civilization.**
>
> It is a **transitional structure**.
>
> AI now makes it possible—for the first time in history—
> to reorganize society not around **how to educate people**,
> but around **how not to obstruct human growth**.
>
> This is the foundation of the next civilizational form:

Human Growth Center

Not an upgrade of education,
but what comes **after** education.

Not chaos—
but humanity's first mature system
built on **trust, reality, and responsibility**.

Part II — The Manifesto of Human Growth in the Age of AI

The Manifesto of Human Growth in the Age of AI

Preface | Why Education Must End

We are not here to reform education.
We are not here to upgrade education.
We are not here to "empower education with AI."

We are here to end education.

Not because education has failed,
but because it has **completed its historical mission**.

I. Education Has Never Been Neutral

Throughout most of human history, "education" has never been a
natural or innocent concept.
From its very beginning, it has been a **structural instrument of
civilization**.

Education has been used to:
- Standardize language
- Enforce obedience
- Train functional skills
- Sort populations
- Allocate social positions
- Stabilize political order

In agricultural societies, education served lineage and
hierarchy.
In industrial societies, it served efficiency and division of
labor.
In the age of the nation-state, it served identity, legitimacy,
and governance.

Education was never designed for human self-realization.

Its core purpose was always singular:

To make human beings fit into a pre-designed social structure.

II. Why Education Was Inevitable Before AI

One historical fact must be acknowledged honestly:

Before the emergence of AI,
education was the only scalable mechanism for transmitting human capability.

Because in that era:
- Knowledge was scarce
- Information was expensive
- Teachers were the sole interface
- Schools were the sole infrastructure
- Credentials were the sole signal

Humanity had no alternative but to accept:
- Standardized curricula
- Age-based grouping
- Linear progression
- Early selection
- Fixed life trajectories

Education was imperfect, often violent in its effects.
But it was, at the time, **the only structure that could function**.

III. AI Ends Education's Historical Necessity

AI does not represent "better educational tools."
It represents a **structural rupture**.

For the first time in history, AI simultaneously dissolves the three foundations upon which education depended:

1 Knowledge Is No Longer Scarce

Explanation, demonstration, iteration, feedback, and synthesis are no longer monopolized by teachers or institutions.

2 Learning No Longer Requires Institutions

Learning can occur at any moment, in any place, at any age, across an entire lifetime.

3 Capability No Longer Requires Credentials

Real competence, creative output, and growth trajectories are becoming more informative than certificates or degrees.

When these three conditions exist simultaneously, one conclusion becomes unavoidable:

> **Education, as a system centered on transmission, filtering, and certification,**
> **has lost its structural necessity.**

IV. Why "AI-Enhanced Education" Is a Civilizational Mistake

The world is currently making a profound error:

Attempting to **rescue education using AI**.
- Using AI to accelerate test preparation
- Using AI to intensify surveillance and assessment
- Using AI to optimize early selection
- Using AI to automate ranking and evaluation

This is not progress.
It is **technological life support for a system that should be allowed to end**.

The outcome will be:
- Earlier selection
- Deeper inequality
- More opaque control
- A more complete reduction of humans into data objects

When a system is no longer necessary but becomes infinitely amplified,
it turns from an institution into a cage.

V. We Are Not Ending Learning — We Are Ending "Being Educated"

This distinction is essential.

We do not oppose:
- Learning
- Growth
- Exploration
- Practice
- Knowledge
- Skill acquisition

What we reject is:

A system built on the assumption that humans must be shaped, managed, and sorted.

What must end is:
- The division of life into "learning years" and "usable years"
- The illusion that age 18 determines destiny
- The replacement of human value with scores, rankings, and credentials

Learning will not disappear.
Only the legitimacy of education as a governance system will.

VI. From Education to Human Growth: A Civilizational Shift

For the first time, AI allows humanity to ask a question that was previously impossible to realize:

**If we no longer manage people through education,
how should society be organized?**

Our answer is clear:

Human Growth must replace Education as the core organizing principle of civilization.

This means:
- Humans are not products to be standardized
- Growth paths must not be designed around selection
- Failure is not a defect
- Happiness must not be postponed until compliance is complete

Instead, we acknowledge:
- Growth is nonlinear
- Awakening can occur at any age
- Human value is not comparative
- Society's responsibility is to avoid blocking growth

VII. This Is Not Reform — It Is a Civilizational Decision

Ending education is not a policy proposal.
It is not a national reform plan.
It is not a technological utopia.

It is a **civilizational judgment**:

125

When humanity finally possesses the capability to support growth without coercion or sorting, continuing to rely on education becomes morally indefensible.

Conclusion | The Meaning of This Manifesto

This is not a manifesto against teachers.
It is not a manifesto against schools.
It is not a manifesto against effort.

It is a declaration that:

Human beings no longer need to be shaped in order to justify their existence.

For the first time in history, AI allows us to say — and to mean — this sentence:

A human life is not educated into existence. It is allowed to grow into itself.

Chapter 11 | How a Human Life Grows: A Continuous Trajectory from Childhood to Old Age

The industrial age did not only transform production.
It fundamentally reshaped how humanity understands **life itself**.

Under industrial logic, a human life was forcibly divided into several isolated, function-defined stages:
- **Childhood**: to be shaped
- **Youth**: to be selected
- **Midlife**: to be used
- **Old age**: to be exited

This segmentation appears natural.
In reality, it has nothing to do with life.

It did not arise from an understanding of human nature,
but from the needs of **system management, efficiency control, and social stability**.

When society demands predictable labor, orderly progression, and low governance risk,
life is redesigned into a linear pipeline:

prepare → use → discard

From the perspective of **Human Growth**,
this interpretation is itself a fundamental mistake.

A human life is not a sequence of stages stitched together.
It is a continuous, evolving process of self-growth.

I. Life Is Not a "Prepare–Use–Discard" Process

To define life as a process
is to reduce the human being to an instrument.

This logic quietly assumes that:
- Only certain periods of life are "valuable"
- Other periods are merely transitional or residual
- Human worth depends on whether the system still needs you

Within this structure:
- Childhood is "not yet ready"
- Youth must stabilize quickly
- Midlife must continuously produce
- Old age becomes a burden

But a real human life has never been a project serving a system.

A human life is an ongoing process of self-understanding, self-adjustment, and self-growth.

Value does not come from being used.
It comes from **continuing to grow**.

II. Childhood: Not to Be Shaped, but to Be Protected

In the old system,
childhood is defined as an "incomplete state."

As a result, control is easily justified:
- Correction
- Standardization
- Comparison
- Uniformity

But in a **Human Growth–centered** framework,
the core task of childhood is singular:

To protect the conditions of growth.

More precisely:

128

To protect curiosity from being prematurely worn down by systems.

Childhood does not need acceleration.
It does not need life planning.

What it truly needs is:
- A sense of safety
- Space for free exploration
- An environment without ranking
- Companionship without evaluation

A person who loses curiosity in childhood
often spends a lifetime trying to recover it.

III. Youth: Not to Be Selected, but Allowed to Try and Fail

In industrial education narratives,
youth is compressed into a one-way channel:
- Choose the right major
- Enter the right institution
- Lock into the correct track

Deviation is labeled failure.

But in a real human life,
the core value of youth lies precisely in **uncertainty**.

In a Human Growth framework,
youth is redefined as:

A period where extensive, real, and reversible trial and error is permitted.

- Direction may change
- Interests may pause
- Identity may be rebuilt

Youth does not need premature stability.

Stability imposed too early
creates instability for a lifetime.

IV. Midlife: Not to Be Used, but to Enter Deep Growth

In the old narrative,
midlife is synonymous with:
- Responsibility
- Exhaustion
- Pressure

One must continuously prove one is "still useful."

But in real life trajectories,
midlife is often a critical turning point:
- Experience begins to consolidate
- Judgment becomes mature
- Ability aligns with inner values

This is the **golden period of deep self-growth**,
not a phase for extraction.

When society stops defining value purely by output,
midlife no longer becomes a source of anxiety,
but a beginning of integration and clarity.

V. Old Age: Not Exit, but Role Transformation

The industrial age could not understand old age,
so it chose to marginalize it.

From a Human Growth perspective, however,
old age is not defined by decline,
but by **role transformation**.

130

From:
- Executor → Observer
- Competitor → Witness
- Self-proof → Understanding others

This is the stage most suited to becoming:
a transmitter of meaning and a guardian of values.

A society that cannot place old age
is a society that cannot understand a complete human life.

VI. What Happens When Life Is Understood as Continuous Growth

Once society stops slicing life into rigid stages,
long-standing anxieties naturally dissolve:
- Fear of being "too late"
- Obsession with "missing the window"
- Self-negation based on age

People realize, for the first time:

As long as life continues, self-growth continues.

Life is no longer a countdown.
It becomes a process of continuous realignment.

VII. Intergenerational Coexistence, Not Segregation

Stage-based life inevitably produces generational separation:
- The young are "immature"
- The middle-aged are "pressure sources"
- The elderly are "burdens"

In a continuous Human Growth perspective:
- Different ages represent different densities of growth

131

- Every stage carries irreplaceable value
- True wisdom emerges from intergenerational dialogue

A healthy society is not one of generational competition, but one of **mutual nourishment across ages**.

VIII. Life Does Not Need to Be "Completed" — Only Aligned

The old system is obsessed with completion:
- Academic completion
- Career completion
- Goal completion

But life has never been a checklist.

From the perspective of Human Growth, life does not need completion.

It needs only one thing:

Continuous alignment with one's potential, judgment, and inner conscience.

Chapter Conclusion

When life is re-understood as a continuous process of self-growth, people no longer rush to prove:

"Do I still have value right now?"

Because value no longer belongs to a single stage, but to a **life that continues to grow**.

Chapter 12 | Education Enters the Museum of History

Every true leap in human civilization
is accompanied by the disappearance of a word.

Not because the word was wrong,
but because it completed its mission.

Today, we must acknowledge a fact that is difficult, but
unavoidable:

The word "education" no longer belongs to the future.

I. Education Was Never Designed for Human Beings

Education is not a natural or eternal institution.
It emerged under very specific historical conditions:
- Knowledge was scarce
- Information was closed and slow
- Societies needed people shaped into predictable, manageable forms

Thus, across different stages of civilization:
- **In agrarian societies**, education served order and inheritance
- **In industrial societies**, it served discipline, standardization, and obedience
- **In early information societies**, it served selection, competition, and efficiency

Yet in all these stages, education rarely served the human being as such.

It was primarily concerned with:
- What kind of people the state needed
- What kind of people the market required

- How systems could remain stable and controllable

It almost never asked:
- Who is this person?
- What allows this person to live with dignity and happiness?
- Is this person's potential being quietly wasted?

Education's true function was never "helping people become themselves."
It was about **producing people suitable for the system**.

II. AI Collapses the Foundations of Education All at Once

Education depended on three long-standing assumptions:

Assumption One: Knowledge Holders Are Scarce

Teachers know; students do not.

Assumption Two: The Correct Path Is Known

Curriculum, grades, schedules, standard answers.

Assumption Three: Humans Must Be Trained to Be "Useful"

Employable, competitive, measurable.

AI dismantles all three simultaneously:
- Knowledge is no longer scarce
- Paths are no longer singular or fixed
- "Usefulness" is no longer the highest human value

When any child can access—within seconds—
a density of knowledge and analysis that exceeds any single teacher,

maintaining the *form* of education
is no longer backwardness alone—

it becomes a **systemic illusion**.

III. The Ox Cart on the Airport Runway

Today's education systems appear more "advanced" than ever:
- Denser curricula
- More homework
- More frequent testing
- More technology
- Deeper anxiety

Yet structurally, they resemble a well-maintained ox cart rolling carefully across a runway designed for aircraft.

The problem is not effort.
It is not intelligence.
It is not intention.

The problem is simple:

The system belongs to a previous civilization.

IV. Why "Education Reform" Is the Wrong Question

We often hear:
- Education reform
- Education modernization
- Education upgrading

All these phrases hide a dangerous assumption:

That education itself is fundamentally correct—
it merely needs improvement.

135

This is false.

You do not "reform" a carriage
to make it compete with high-speed rail.

When civilization's foundations shift,
what must be replaced is not method—but **concept**.

V. Humans No Longer Need to Be Educated

This statement sounds radical.
In the age of AI, it is remarkably simple:

Humans no longer need to be educated.

Because:
- Knowledge is instantly accessible
- Skills can be acquired on demand
- Information is no longer a life barrier

What is truly scarce is not:
- What you know

But rather:
- What you care about
- What direction you are willing to commit your life to
- Whether you can sustain self-directed growth
- Whether you can align your actions with judgment, values, and conscience

These cannot be taught.

VI. Education's End Is Not Collapse—It Is Completion

Placing education in the museum of history
is not a rejection.

It is recognition.

Just as we do not mock:
- •   Farming tools
- •   Steam engines
- •   Typewriters

They once moved civilization forward.
They simply no longer belong to the future.

Education is no different.

VII. The New Civilizational Function: Human Growth

When education steps aside,
society does not fall into chaos.

A new function emerges.

This function:
- •   Does not shape people
- •   Does not train people
- •   Does not rank or filter people

It does only one thing:

It creates the conditions for every individual's self-directed growth.

This is the foundation of a new civilization.

We call it:

Human Growth

VIII. From This Chapter On, Language Must Change

Language is not cosmetic.
It reveals civilization's structure.

From this chapter onward:
- We no longer speak of "students"
- We no longer center "teachers" as authorities
- We no longer organize life around curricula or grades

Instead, we speak of:
- Individuals in continuous growth
- Guides and companions
- Open growth environments
- Trajectories that cannot be quantified

Changing language is the first real step toward changing civilization.

Conclusion

When education completes its historical role,
human beings are finally released from being instruments.

From this moment forward,
the human being returns to what it should always have been:

the purpose, not the means.

Chapter 13 | After the End of Education: A New Human-Centered Civilization

When the word *"education"* completes its historical mission,
the world does not fall into emptiness.

On the contrary—
a civilizational core long obscured
emerges clearly for the first time.

I. The End of Education Is Not the Beginning of Chaos, but a Sign of Civilizational Maturity

Human societies once needed "education"
not because human beings were incapable of learning,
but because **systems fundamentally distrusted people**.

They distrusted:
- human judgment
- human rhythm
- human goodwill
- human capacity for self-growth

As a result, societies invented mechanisms to replace trust:
- standards
- curricula
- evaluation
- ranking

These mechanisms once served a purpose.
They stabilized early societies, replicated skills, and preserved order.

But in the age of AI, a truth becomes unavoidable:

Education is, at its core, a system built on distrust of human beings.

When a society finally has the capacity—and the courage—to let go of that distrust,
the end of education ceases to be a danger signal
and becomes a marker of civilizational adulthood.

II. Human-Centered Does Not Mean "Centered on the Successful"

One long-standing confusion must be clarified.

"Human-centered" does **not** mean:
- elite-centered
- talent-centered
- efficiency-centered
- winner-centered

A truly human-centered civilization means:

No person is required to prove their right to exist.

From the moment of birth,
every human being already possesses full dignity and legitimacy.

The task of civilization is not to select "worthy" people,
but to ensure that **no system systematically crushes, filters out, or negates anyone**.

III. When Civilization Shifts from "Efficiency" to Human Growth

The supreme value of industrial civilization was efficiency.

Efficiency presupposes:
- singular goals

- fixed pathways
- replaceable humans

But human beings have never fit these conditions.

In a civilization centered on **Human Growth**,
the foundational objective shifts irreversibly:
- from "How fast?"

to "Did it genuinely happen?"
- from "Maximum output"

to "Was human potential actually used?"
- from "Conformity to standards"

to "Alignment with one's inner direction."

Efficiency no longer governs civilization.
It returns to its proper place—as a tool, not a ruler.

IV. The Minimum Ethics of Future Institutions: Non-Interference

A future civilization does not require perfect institutions.
It requires only one principle—
extremely restrained, yet extraordinarily difficult:

Do not obstruct.
- Do not obstruct exploration
- Do not obstruct change
- Do not obstruct lateness
- Do not obstruct failure
- Do not obstruct difference

This may sound like inaction.
In reality, it demands exceptional institutional discipline.

It requires systems capable of **not rushing to correct,
not rushing to judge,
not rushing to force people back into "the right place."**

141

V. In the Age of AI, Humanity Returns to the Human Question

As AI takes over:
- knowledge acquisition
- computation
- prediction
- optimization

human beings are forced to confront a question that cannot be outsourced:

Why do I exist?

This is no longer a philosopher's question alone.
It becomes an everyday reality for ordinary people.

A de-educationized society does not answer this question for you.
But it guarantees one essential condition:

The system will not answer it in your place.

VI. The Fundamental Unit of the New Civilization: The Growing Individual

In the old civilization, society recognized units such as:
- labor power
- students
- professional identities
- utility labels

In the new civilization, only one unit is acknowledged:

A human being in continuous self-growth.

This person:
- may change direction

142

- may pause
- may repeat
- may remain silent
- may begin again

Civilizational quality is no longer measured by GDP, rankings, or speed,
but by a single, profound question:

How many people, in their daily lives, genuinely feel they are becoming themselves?

VII. This Is Not Utopia—It Is the Only Sustainable Path

Some will say:
"This sounds idealistic."

But what is truly unrealistic
is maintaining the old system.

Because:
- suppressed potential becomes social cost
- denied difference becomes conflict
- perpetually evaluated people become numb,

angry, or empty

A society that blocks human growth
cannot survive in a technologically advanced world.

VIII. The Responsibility of Our Generation

Every generation encounters an unavoidable civilizational threshold.

For us, that threshold is this:

When technology has already liberated knowledge,

are we willing to liberate one another?

Ending education is not a rejection of the past.
It is an acknowledgment that:

The past has completed its task.

Final Declaration

Education is not the future.
Human beings are.

When society stops trying to shape, manage, and filter people,
and instead makes room for every individual's self-growth,

human civilization
finally reaches maturity.

Chapter 14 | The Fundamental Rupture Brought by AI

AI Is Not a Tool Upgrade — It Is a Civilizational Break

Throughout history, humanity has experienced many technological advances.
Most of them, however, were merely **tool upgrades**:
- Faster machines
- More efficient processes
- More refined systems of management

They improved productivity,
but they did **not** fundamentally challenge one central assumption:

What does learning actually mean?

AI does.

For the first time, technology does not merely improve learning efficiency.
It **breaks the meaning of learning itself**.

This is not an upgrade.
It is a rupture.

I. The Industrial Logic of Learning Is Collapsing All at Once

Education systems of the industrial and early information eras were built on a stable set of assumptions:
- Knowledge is scarce
- Information is expensive
- Correct answers are limited
- Skills require long-term training
- Learning outcomes must be verified, ranked, and filtered

From these assumptions emerged a full institutional structure:

- Classrooms
- Curricula
- Exams
- Credentials
- Rankings

For a long time, this structure was rational.

AI causes **all of these premises to collapse simultaneously**.

II. Five Fatal Reconstructions of Learning by AI

1 Knowledge Is No Longer Scarce

Before AI, the primary barrier to learning was:

"I cannot access high-quality knowledge."

Today:
- High-level explanations are instantly available
- Multiple perspectives are accessible at once
- World-class knowledge is no longer tied to elite institutions

Knowledge is no longer a gate — it is a background condition.

2 Memory Is No Longer Central

Memory was once the core of learning:
- Memorization
- Recall
- Rapid retrieval

AI surpasses humans completely in:
- Memory
- Search
- Comparison
- Pattern recognition

Continuing to treat "how much you remember" as competence
has lost all technical justification.

3 Skills Can Be Instantly Augmented

Previously, skills meant:
- Long-term training
- Repetition
- Accumulated mastery

Now:
- Coding can be assisted in real time
- Writing can be instantly refined
- Design can be co-generated
- Analysis can be dynamically enhanced

Skills are no longer about possession,
but about **invocation, orchestration, and recombination**.

4 Correct Answers Are Everywhere

Standardized answers once formed the authority of education.

In the AI era:
- Multiple valid answers coexist
- Problems themselves can be redefined
- Answers shift with context

The idea of a "single correct answer"
has collapsed at the technological level.

5 Learning Is No Longer Linear

Traditional education assumed:
> • Age → grade → curriculum → exam → qualification

AI enables:
- Nonlinear learning
- Instant jumps
- Cross-domain synthesis
- On-demand knowledge creation

Learning is no longer a pre-life phase.
It becomes an **always-on human activity**.

III. When Everything That Can Be Taught Can Be Automated

A sober observation reveals a harsh truth:

**Almost everything that can be "taught"
can now be replaced, assisted, or surpassed by AI.**

Including:
- Knowledge
- Techniques
- Methods
- Standardized procedures

So what remains?

IV. What Remains Is the Human Being

In the age of AI, what cannot be replaced belongs to only one category:

Human capacities.

These are not technical abilities.
They are existential ones.

- **Judgment**
  - Choosing under incomplete information
  - Weighing conflicting values

- **Sense of Value**
  - Knowing what matters
  - Knowing what does not

- **Imagination**
  - Conceiving what does not yet exist
  - Opening new possibilities

- **Empathy**
  - Understanding lived experience
  - Sensing non-quantifiable pain and hope

- **Directional Choice**
  - Not "what can be done," but "why it should be done"

- **Care for Others and the World**
  - Not optimizing systems
  - But caring for life itself

These capacities:
- Cannot be tested
- Cannot be standardized
- Cannot be mass-produced

They can only be **protected, nurtured, and allowed to emerge**.

V. Defining Humans by Exams Is Now a Civilizational Error

In the AI era, continuing to define people by scores, rankings, and standardized answers is no longer just outdated.

It is:

A systematic waste of humanity.

It wastes:
- Judgment
- Imagination
- Emotional intelligence
- Late bloomers
- Nonlinear lives

This is not a policy mistake.
It is a **civilizational misjudgment**.

VI. The Real Rupture Is Not Technological, but Anthropological

The greatest impact of AI is not how intelligent machines become.

It is the question AI forces upon humanity:

If machines now learn better than we do, why should humans learn at all?

The answer can no longer be:
- For exams
- For credentials
- For selection

The only remaining answer is:

To become fully human.

VII. From a "Learning Society" to a "Human Growth Society"

The true transition of the AI era is:
- From teaching people
- To **not obstructing human growth**
- From managing learning
- To **supporting self-growth**
- From selecting the few
- To **releasing the many**

Learning is no longer the center of civilization.
Human beings are.

Chapter Conclusion | Only After the Rupture Does Direction Become Clear

AI does not diminish human value.

It merely reveals, with brutal clarity:

Which things never belonged to the core of human value in the first place.

Once knowledge, memory, skills, and answers are liberated by technology,
humanity is finally forced to confront a long-avoided question:

> Who am I?
> What do I care about?
> What am I willing to give to the world?

> AI cannot answer these questions.

> But any civilization worthy of the future
> must preserve them
> for every human being.

Chapter 15 | The True Role of AI: Not a Teacher, but an Amplifier of Human Potential

When AI enters the learning domain,
the most common—and most dangerous—mistake people make is this:

They treat AI as a better teacher.

This misunderstanding is not harmless.

If AI is merely used to
explain more clearly,
answer more quickly,
grade more accurately,

then humanity has done nothing but upgrade the old education system
into a **more efficient machine of control**.

That is not progress.
It is acceleration in the wrong direction.

I. AI Did Not Emerge to Teach Humans

AI was not created for "education."

Its fundamental capabilities are:
- ultra-fast information retrieval
- parallel processing of complex systems
- simulation of countless possibilities
- continuous, responsive dialogue

These capabilities do not point toward *teaching*.
They point toward **amplifying exploration**.

The role of "teacher" presupposes:

- authority
- correct paths
- predefined outcomes

In the age of AI,
this role is not only unnecessary—
it is structurally obsolete.

AI is poorly suited to be a teacher
because *teaching*, as an institutional role,
is what must disappear.

II. AI's Correct Position: An Amplifier of Human Potential

In a civilization centered on **self-growth**,
AI must be redefined as:

An Amplifier of Human Potential.

Its function is not to replace human thinking,
but to **release capacities long suppressed by systems**.

For example:
- Curiosity is supported instead of interrupted by schedules
- Understanding of complex systems is no longer limited by years of prerequisite training
- Creative attempts are no longer abandoned due to technical barriers

AI does not push people onto paths.
It **widens the space in which paths can be discovered**.

III. AI Does Not Eliminate Effort—It Eliminates Artificial Barriers

The cruelty of traditional education lies in this illusion:

It disguises *barriers* as *effort*.
- memorization burdens
- endless repetition
- meaningless foundational stacking

These were never necessary routes to wisdom.
They were **filtering devices of the industrial age**.

AI's true value is not that it makes people lazy.

Its value lies in this:

It removes gates that never should have existed, allowing people to reach judgment, synthesis, and creation directly.

Effort remains.
But it is finally directed toward what matters.

IV. When Everyone Has a Personal AI Companion

In the new paradigm,
every individual naturally possesses a:

Personal AI Companion.

This is not:
- a surveillance tool
- a scoring system
- a learning management platform

It is a persistent dialogic presence used to:
- explore interests

- expand perspectives
- simulate decisions
- examine consequences
- challenge one's own assumptions

AI does not tell you what to do.
It helps you see **what you could not see before**.

V. What AI Can Never Replace Reveals the Human Core

The stronger AI becomes,
the clearer one truth emerges:

Human value was never about speed, accuracy, or storage.

What AI can never replace is:
- value judgment
- meaning selection
- ethical responsibility
- sensitivity to others' lived realities
- answering the question: *Why should this be done at all?*

If a society hands these questions to AI,
it is not advancing.

It is abandoning its humanity.

VI. The Stable Structure: AI × Guide × Human

In a self-growth-centered civilization,
the most stable and humane structure is triangular:
- **AI** expands possibilities
- **Guide** safeguards direction and ethical grounding
- **Human** makes the final choices

When one element dominates, imbalance follows:
- AI alone → technocracy
- Guide alone → human authority and control
- Individual alone → isolation and disorientation

Balance is not optional.
It is essential.

VII. Why "Using AI to Teach Better" Is the Wrong Question

"Improving teaching efficiency with AI"
sounds progressive.

In reality, it merely delays collapse.

It assumes three false premises:
1. Humans still need to be taught
2. Content remains the core
3. Evaluation is still legitimate

AI is proving that **none of these are true**.

The problem is not that education is inefficient.
The problem is that **education itself no longer fits reality**.

VIII. The Real Challenge of the AI Age Is Not Technology, but Courage

The technology is ready.

What is missing is courage:
- the courage to relinquish control
- the courage to abandon ranking
- the courage to let go of the arrogance that says
"I know how you should live."

AI is not here to save education.

It is here to **force humanity to confront its deep mistrust of people**.

Chapter Closing

AI has not made humans unnecessary.

It has merely stripped away
the institutional shells we used
to hide our distrust of human beings.

As AI takes over knowledge,
humanity is finally free
to return to a question long avoided:

Who am I?
What am I willing to take responsibility for?

That question cannot be taught.
But a worthy civilization
must never prevent anyone from asking it.

Chapter 16 | Redefining a Happy Life

Happiness Is Not Success — It Is the Full Release of Human Potential

In human society, the word **"success"** has been used for so long that people have almost forgotten to ask a more fundamental question:

What is success for?

Wealth, status, fame, influence—
the things repeatedly implanted as "life goals"—are, in essence, nothing more than **external coordinates inside a social evaluation system**.

They measure:
where you stand in comparison,

not:
whether you have truly become yourself.

I. Happiness Has Never Been the Result of Comparison

A person can be highly successful
yet feel, deep inside, hollow, anxious, and exhausted.

Another person
may have no prestigious labels,
yet feel every day a quiet sense of fullness, focus, and peace.

This is not merely a difference in personality.
It is a collective illusion created by **a wrong evaluation system**.

When a society defines happiness through "success,"
it inevitably forces people into constant comparison;
and once comparison becomes the core logic,

most people are mathematically destined to be defined as "failures."

This is not a moral issue.
It is a mathematical fact.

II. The Systematic Misdirection of "Happiness" in Old Civilization

Industrial civilization and institutional society
quietly replaced happiness with a set of indicators that are
manageable, comparable, and rankable:
- wealth
- social status
- job titles
- recognition from others
- "winning" in competition

These indicators share one defining feature:

They are external, relative, and defined by other people.

So people were trained to understand life like this:
- **Happiness = being better than others**
- **Success = not being eliminated**
- **Value = being needed by the system**

In this narrative, happiness stops being an inner experience
and becomes an endless game of comparison.

The result is predictable:
- winners do not dare to stop
- losers cannot forgive themselves
- everyone moves forward in fear

This is not progress.
It is the **systematic consumption of human beings**.

III. Happiness Is Not a "Result" — It Is a Continuous State

From the perspective of **Human Growth**,
happiness is not an endpoint,
not a prize,
not something granted after "achievement."

Happiness is a **living state**—continuously occurring in life.

It emerges through a process in which:

a person,
without being forced, distorted, or suppressed,
continues to use, align, and develop their unique potential.

Happiness is not "what I have completed."
It is:
- whether I am truly living
- whether I am using the abilities that belong to me
- whether my actions are aligned with my inner motives
- whether I feel life genuinely happening

This is an inner intensity,
not external certification.

IV. The Real Source of Happiness: Potential in Use

Happiness is not a brief emotion.
It is a stable experience of existence.

When does it appear?

Not when you "surpass others,"
but when you:
- are using your most natural abilities
- feel focused, rather than torn apart

- act in strong alignment with inner motivation
- clearly sense: **"This is what I am meant to do."**

Psychology often describes this as "flow,"
but more deeply, it is:

the feeling that life is allowed to run as it truly is.

Unused potential becomes anxiety.
Suppressed potential becomes depression.
Only potential that is continuously used
transforms into happiness.

V. Humans Are Born Different: This Is a Fact, Not a Problem

Almost every traditional education and social system
avoids an obvious truth:

Human beings are born with profoundly different structures of
potential.

Different ways of perceiving,
different rhythms,
different densities of interest,
different emotional sensitivity,
different paths of attention and focus.

But industrial society could not process difference,
so it chose the simplest—and most violent—solution:

treat difference as deviation.

Unified curriculum, unified pace, unified standards, unified
evaluation—
this is not education.
It is the systemic denial of difference.

The real tragedy is not difference itself,
but this:

measuring every life with the same ruler.

VI. What Happens When Potential Is Suppressed Long-Term

A society that continuously suppresses individual potential
will inevitably pay three long-term prices:

1) Creativity dries up
People stop trying; they only try not to be wrong.

2) Psychological systems collapse
Anxiety, depression, and emptiness become normal.

3) Meaning systems disintegrate
People do not know why they live; they only know they must not
fail.

These problems will not be solved through "better education
reform,"
because they are produced by the **logic of old education and old
success itself**.

VII. Happiness Requires a Complete Break from the "Comparison Society"

Comparison is the core tool through which industrial civilization
maintains order.
But in a civilization of Human Growth, comparison must exit.

The reason is simple:
- comparison always points outward, toward others

- happiness always happens inward, within the individual

A person who constantly compares
cannot perceive their own real state.

Therefore, redefining happiness necessarily means:
- exiting rankings
- exiting templates
- exiting the "correct life" defined by others

This is not avoidance.
It is a return to life itself.

VIII. Realigning Happiness with Time

Another deep distortion of happiness in old civilization
is its manipulation of time:
- endure now
- you'll be happy later
- just hold on a few more years
- wait until conditions are ready

But reality is often this:

happiness is postponed repeatedly
until life is exhausted.

In the Human Growth perspective,
happiness is not "future compensation."
It is:

whether, in every stage of life,
you allow yourself to be the person who is still growing.

Happiness does not require "completion."
It requires continuous alignment.

IX. In the Age of AI, Happiness Becomes "Accessible" for the First Time

Before AI, happiness was a luxury for most people, because:
* resources were scarce
* opportunities were concentrated
* learning costs were extremely high
* the price of trial and error was enormous

AI changes one condition at the technical level for the first time:

everyone may continuously grow in their own way.

When:
* knowledge is no longer a barrier
* skills can be obtained on demand
* creation no longer requires a gatekeeping ticket

happiness stops being the privilege of a few
and becomes a widespread possibility.

X. The Minimum Condition for Happiness: Not Being Systemically Blocked

Happiness does not require perfect institutions.
It only requires one minimum condition:

the system must not block people.
* do not block exploration
* do not block change
* do not block being "late"
* do not block failure
* do not block being different

When society achieves this,
happiness does not need to be "designed."
It emerges naturally.

Closing | Happiness Is Not Given — It Is Allowed

Happiness is not a reward.
It is a signal.

It tells you:
you are living on a path aligned with your potential.

A happy life is not a pre-planned route of success.
It is this:

a person, across an entire lifetime,
is not forced to give up the right to become themselves.

When civilization no longer demands that people prove they
"deserve happiness,"
happiness returns to human hands.

Happiness is not the result of comparison.
It is the feeling that Human Growth is truly happening.

Chapter 17 | Abolishing Evaluation: Why Human Beings Should Not Be Measured

If examinations are the most visible form of violence in the old education system,
then **evaluation** is its most hidden—and most enduring—mechanism of control.

Scores, grades, rankings, GPAs, comments, labels—
they appear gentle, rational, professional.
Yet quietly, persistently, they transform *human beings*
into objects that can be compared, ordered, and managed.

Evaluation does not shout.
But it shapes an entire life.

I. The Fundamental Assumption of Evaluation Is Wrong

All systems of evaluation rest on an assumption that is almost never truly examined,
yet is silently accepted by modern society:

That a person's value can be accurately measured by external standards.

Reality proves the opposite:
- Human potential is dynamic
- Human states are fluid
- Human directions are non-linear
- Human maturity is unpredictable

Using fixed scales to measure a living, changing human being
is not an issue of *inaccuracy*—
it is a **conceptual error**.

The problem with evaluation is not that it fails to measure well.
The problem is that it attempts something that is fundamentally impossible.

II. Evaluation Is Never Descriptive—It Is Formative

Evaluation is never merely "recording facts."

What it actually does is **reshape behavior in reverse**.

When a person lives inside an evaluation structure for long enough,
they begin to ask instinctively:
- What will be approved?
- What will be penalized?
- What is the "safe" choice?
- What should be displayed, and what must be hidden?

As a result:
- Exploration gives way to strategy
- Authenticity gives way to performance
- Inner motivation gives way to external feedback

For every day evaluation exists,
a person has one less day of being themselves.

III. The Deepest Harm of Evaluation: Outsourcing Judgment

The true danger of evaluation is not receiving a low score.
It is the **systematic outsourcing of judgment**.

When people grow accustomed to being measured,
they gradually lose a critical human capacity:

The ability to judge whether they are living a real, complete, meaningful life.

They stop asking:

"Am I fulfilled?"

And start asking:

"How do others see me?"

This is not failure.
It is the slow erosion of personhood.

IV. Why "Kind" or "Positive" Evaluation Is Still Harmful

A common objection arises:

"Evaluation doesn't have to be oppressive.
It can be encouraging."

But the problem is structural.

As long as evaluation exists, **power asymmetry exists**.

Even "positive feedback" carries a hidden message:

"You are being watched, approved, permitted."

This subtly shifts motivation from the inside back to the outside.

The moment a person acts in order to be approved, self-growth has already been interrupted.

V. In the Age of AI, Evaluation Loses All Technical Necessity

In the old world, evaluation solved a practical problem:

Resources were scarce, so selection was required.

In the age of AI, this premise collapses simultaneously:
- Knowledge is no longer scarce
- Tools are no longer expensive
- Creative paths are highly diverse
- Value is no longer concentrated in a single role or skill

Selection is no longer a prerequisite for social functioning.

Evaluation therefore reveals its true identity for the first time:

It is not a necessary tool—
it is a **historical residue of control habits**.

VI. Without Evaluation, How Do People Grow?

This is the most common—and most sincere—question.

The answer is simple:

Through feedback, not evaluation.

The difference is fundamental:
- **Evaluation** assigns you a position
- **Feedback** helps you understand a process

Feedback focuses on:
- What happened
- What options existed
- What different choices might lead to
- Where the next step could go

169

It does not define who you are.
It helps you understand what you are doing.

VII. The Only Record That Remains: Life Trajectories

In the **Human Growth** framework,
only one form of record is preserved:

The Life Exploration Archive

It is not a transcript.
Not a résumé.
Not a certificate of ability.

It records only:
- Personal exploration journals
- Traces of projects and attempts
- The evolution of thinking
- Reasons behind key choices
- Failures, pivots, and restarts

It is never used for comparison.
Only for reflection.

VIII. What Happens When People Are No Longer Measured?

When evaluation systems disappear,
people pass through a brief but necessary phase:
- Anxiety
- Disorientation
- Loss of external reference points

Then deeper changes emerge:

1 **Motivation Returns Inward**

People begin doing *meaningful* things,
not merely things that *look useful*.

2 **Failure Loses Its Shame**

Failure becomes part of exploration again,
not a negation of identity.

3 **Human Relationships Change**

People stop being competitors,
and begin becoming companions.

IX. A Society Without Evaluation Will Not Collapse

On the contrary.

True disorder comes from forcing large numbers of people
to advance along paths that do not fit them.

When people are allowed not to be measured,
society finally sees:

The real distribution of human potential—
not the distorted version produced by evaluation systems.

X. Evaluation as a "Low-Cost Governance Technology"

Evaluation is used globally not because it is scientific,
but because it is **cheap and efficient**.

For institutions, evaluation means:
- No need to understand people
- No need to accompany growth
- No need to take long-term responsibility

- Only metrics, data, and conclusions

Evaluation replaces understanding with numbers.

It compresses complex human reality into manageable tables,
and shifts responsibility from systems to individuals.

Once you are labeled "low-scoring" or "unqualified,"
the question is no longer whether the system failed—
you failed.

That is its greatest success—and its greatest danger.

XI. The Hidden Alliance Between Evaluation and Capitalism

Evaluation is not neutral.
It is structurally aligned with modern capitalism.

Capital requires:
- Comparable individuals
- Predictable behavior
- Replaceable roles

Evaluation delivers exactly that:
- Ranking → comparison
- Performance → prediction
- Labels → substitution

Under evaluation:
- Humans become "human resources"
- Growth becomes "ability investment"
- Life becomes a "return-on-investment curve"

Once evaluation becomes the default language,
people begin to understand themselves through market logic.

This is not an economic issue.
It is the **colonization of self-perception**.

XII. Why Evaluation Produces "False Elites"

Evaluation does not select the most suitable people.
It selects those who are:
- Best at pleasing rules
- Best at hiding real differences
- Earliest to internalize self-censorship

That is why in highly evaluative systems:
- Truly original individuals are eliminated early
- High achievers fear uncertainty most
- "Excellence" often coincides with fragility

Evaluation does not create strength.
It creates **obedient survivors**.

XIII. Irreversible Harm to Children and Adolescents

For adults, evaluation is pressure.
For children, it is **personality programming**.

When a child repeatedly hears:
- "You are smart / you are not"
- "You have a future / you don't"
- "You are a good student / you are a problem"

They do not learn knowledge.
They learn this:

"My existence requires external permission."

This internalized structure produces lasting effects:
- Unstable self-worth
- Pathological fear of failure

173

- Addiction to external approval

This is not a psychological flaw.
It is **structural trauma caused by evaluation**.

XIV. Why Society Fears a World Without Evaluation

The strongest resistance to abolishing evaluation
does not come from technology,
but from fear:

> *If we stop evaluating, how do we prove we are right?*

Evaluation reassures:
- Managers
- Experts
- Established elites

A non-evaluative society means:
- Authority shifts from judgment to companionship
- Decisions must tolerate uncertainty
- Success loses moral superiority

For many, this is more frightening than inefficiency.

XV. Without Evaluation, How Does Society Build Trust?

The answer is simple:

Shift from hierarchical trust to process-based trust.

Old logic:

> "Who evaluated this person?"

174

New logic:

"What have they done? How do they think? How do they change?"

Trust emerges from:
- Traceable life trajectories
- Real project participation
- Honest records of failure
- Continuous self-correction

This is not chaos.
It is closer to how humans naturally judge one another.

XVI. Abolishing Evaluation ≠ Abolishing Standards

A common misunderstanding:

"Without evaluation, there are no standards."

False.

What disappears:
- External verdicts
- Single scales
- Ranking logic

What remains:
- Real feedback
- Multi-dimensional understanding
- Individual self-alignment

Standards no longer judge *people*.
They help people understand the consequences of choices.

XVII. The End of Evaluation Is the Beginning of an Adult Society

Evaluation treats adults as beings requiring continuous conditioning.

Abolishing it means acknowledging:
- Humans can take responsibility
- Humans can tolerate uncertainty
- Humans can grow without verdicts

This is not permissiveness.
It is **civilizational adulthood**.

Conclusion | Evaluation Is Not Rationality, but the Language of Fear

Evaluation looks rational.
But beneath it lies fear:
- Fear of losing control
- Fear of difference
- Fear of unpredictability
- Fear that humans will not grow as designed

A civilization that abandons evaluation
does not do so because it is perfect,
but because it finally admits:

Human beings are not errors to be corrected.

Humans are not commodities.
They do not require quality inspection.
Humans are not machines.
They do not require performance metrics.

Every system that tries to measure humans
ultimately measures only one thing:

Its own fear and distrust of humanity.

Abolishing evaluation is not abandoning order—
it is choosing, for the first time, to trust people.

Chapter 18 | Abolishing Examinations: The Only Non-Negotiable Starting Point

— And How a Society Without Exams Actually Functions

If the world were allowed to change only **one thing**—
one single change that would truly allow "education" to exit the stage of history—

that change would **not** be curriculum reform,
not teacher training,
not technological upgrades.

It would be this:

The complete abolition of all examinations.

Not optimizing exams.
Not reducing exams.
Not making exams "fairer."

But allowing examinations to **fully withdraw from the core operating structures of human society**.

Because **any system that depends on exams to function** is, by its very nature,
a system that **does not trust people**,
suppresses human potential,
and **maintains order through fear**.

I. Examinations Are Not Neutral Tools — They Are Technologies of Power

For decades, examinations have been described as:
- Objective
- Scientific
- Fair

- Rational

This description is a repeatedly reinforced **social myth**.

Across all civilizations, examinations have never primarily existed to "discover potential."
They have existed to perform **three precise governance functions**:

1. Ranking
2. Filtering
3. Training obedience

Examinations do **not** answer the question:

How do humans grow?

They answer a different question entirely:

How can a system maintain stability at the lowest possible cost?

This is the true reason examinations exist.

II. What Exams Actually Measure — And What They Never Do

No matter how sophisticated the questions appear, examinations ultimately measure only three things:
- The ability to execute instructions under pressure
- The ability to obey standard answers
- The ability to suppress independent judgment within time constraints

These abilities were extremely valuable in the industrial era.

But in the age of AI, they are undergoing **systemic devaluation**.

Because in all three respects:
- AI is faster than humans
- AI is more accurate than humans
- AI is more stable than humans

To continue measuring humans by examinations is to define human value **according to machine advantages**.

This is not a technical error.
It is a **directional error**.

III. "Fair Examinations" Are Logically Impossible

The idea of a "fair exam" rests on a premise that **does not exist**:

That all people should
at the same time,
in the same way,
demonstrate value.

Reality directly contradicts this premise.

Humans differ profoundly in:
- Maturation rhythms
- Interest activation points
- Emotional stability
- Cognitive pathways
- Modes of expression

When real differences exist, unified standards inevitably produce **systemic failure populations**.

This is not a policy flaw.
It is an unavoidable outcome of the design itself.

IV. Exams Do Not Create Failure — They Legalize Incompetence Narratives

The greatest harm of examinations is not elimination.

It is transformation.

They convert *temporary mismatch* into a lifelong internal verdict:

> "I am not capable."

> When people are defined by scores over long periods, they gradually internalize a dangerous self-image:

> "I am fundamentally not good enough."

> Once internalized, this belief persists for life— even if the person later becomes socially "successful."

> A society can survive failure.
> It **cannot** survive mass internalized self-negation.

V. A Historical Truth: In Humanity's Most Critical Moments, Exams Disappear

This is a fact consistently ignored by institutions but repeatedly proven by history:

In humanity's most critical moments, examinations and credentials automatically collapse.

In the following situations, humans never select people through exams:
- Wartime
- Disaster response

181

- Pandemic outbreaks
- Infrastructure collapse
- Emergency scientific and survival breakthroughs

In these moments, society asks only one question:

Who can solve this problem right now?

No one asks:
- What exams you passed
- What certificates you hold
- Whether you are "qualified"

If you can do it, you step forward.
If you cannot, you are replaced immediately.

This is not idealism.
It is **human instinct under survival pressure**.

History's most decisive breakthroughs—
from wartime technologies and vaccine development
to emergency engineering repairs
to open-source security vulnerability fixes—

have overwhelmingly come from:
- Non-standard backgrounds
- Non-credentialed individuals
- Young, cross-disciplinary, and formally
"unqualified" people

They were allowed to try not because institutions were enlightened,
but because **there was no alternative**.

VI. Exams Exist Only in a "Non-Urgent but Distrustful" Era

Examinations are not tools of capability detection.

They are **low-cost substitutes for order**
in periods that are *not urgent* but *deeply distrustful of individuals*.

When civilizations enter truly complex, uncertain, and high-stakes phases,
exams are instinctively abandoned.

The arrival of AI makes something unprecedented possible:

The **judgment logic used only during crises**
can now become **daily, scalable, and sustainable**.

VII. How Does Society Function After Exams Are Abolished?

The answer is **not chaos**.

It is a shift from **credential systems** to **trial systems**.

This structure is called:

The On-Site Trial Period Mechanism

People are not defined first.
Problems are.

VIII. Core Operating Logic of a Post-Exam Society:

Problems × Attempts

Old Model (Exam Society)
- Label people first

- Assign problems afterward

"What level are you?"

New Model (Trial Society)
- Define problems first
- Invite attempts afterward

"Who is willing to try this problem right now?"

Society stops "selecting people"
and instead continuously searches for **real matches between problems and abilities**.

IX. How On-Site Trials Actually Operate (Practically)

1 No background checks — real problems first
- No simulations
- No written tests
- Real problems with real constraints in real environments

2 Clear, limited trial windows
- One week
- One month
- One defined phase

Time itself is the most honest evaluator.

3 Observe process, not scores

Attention is paid to:
- Clarity of thinking
- Speed of adjustment
- Willingness to take responsibility
- Awareness of personal limits

184

- Ability to ask for help

🔲 Success stays, mismatch rotates
- No humiliation
- No permanent labels
- No "failure identity"

Failure is information, not a verdict.

X. Why This Is Actually Safer Than Exams

This is the most counter-intuitive yet crucial point:
- Exams can hide incompetence for years
- Trials expose real ability immediately

Exams are **high-risk selection mechanisms**.
Trials are **low-risk reality filters**.

XI. Where Does Order Come From Without Exams?

Order no longer comes from rankings.

It comes from:
- Real participation
- Responsibility bound to action
- Collaborative dependency
- Traceable action histories

When learning, work, and exploration are embedded in reality itself,
order emerges naturally.

XII. When Exams Disappear, Real Learning Begins

Without exams, people finally begin asking questions that never had space before:
- What genuinely interests me?
- Where do I feel energy rather than exhaustion?
- Which problems am I willing to engage with long-term?

These questions **cannot exist** in an exam-driven system.

Chapter Conclusion | Turning "Exceptional Moments" into Civilizational Norms

Examinations are not being "overthrown."

They are being recognized—through repeated historical proof—as **non-essential structures**.

If exams are not abolished,
all discussion about future learning is merely cosmetic.

Abolishing exams is not radical.

It is the **minimum respect owed to reality**.

When society no longer defines people by examinations,
humanity can finally enter a civilization centered on:
- Self-growth
- Real capability
- Real responsibility

—rather than fear, ranking, and obedience.

On-site Trial Society
The Operating Model of a Post-Exam Civilization

One-sentence core shift

From **"defining people by exams"**
→ to **"discovering capability through real problems"**
→ to **"building social trust through lived action trajectories."**

I. Old Civilization vs. New Civilization: A Core Logic Comparison

▪ Old Civilization (Exam Society)

Dimension | Operating Logic
- **Definition of the person**

Labels, scores, degrees, certificates
- **How people are selected**

Exam first → allocation later
- **Risk structure**

High scores do not guarantee real capability
- **Cost structure**

Low observation cost, extremely high hidden social cost
- **Outcome**

Mass waste of human potential + long-term psychological depletion

▪ New Civilization (On-site Trial Society)

Dimension | Operating Logic
- **Definition of the person**

A continuously unfolding individual
- **How people are selected**

Problem first → trial next
- **Risk structure**

Inability is exposed immediately

187

- **Cost structure**

Continuous observation, but low systemic failure cost
- **Outcome**

Natural distribution of potential + authentic social matching

II. The Core Triangular Structure (Non-separable)   Real Problem

▲
│

Trial Actor ◄──────► Collaborative Environment

All three elements are indispensable.

1 Real Problems

Not exam questions.
Not simulations.

But:
- Real societal needs
- Engineering / research / community / technology / humanitarian challenges
- With constraints, risks, and real-world consequences

2 Trial Actors

Not "candidates," but:
- Individuals willing to take responsibility for a problem
- No credential requirements
- Capability revealed only through action

3 Collaborative Environment (Support + AI)
- AI collaboration systems
- Tools, data, feedback loops
- Guides (not judges, but guardians of direction)

III. The Full Operating Flow of an On-site Trial

[Problem Emerges]
↓
[Problem Clearly Defined]
↓
[Open Call for Trial Participation]
↓
[Short-term On-site Trial]
↓
[Process Recording + AI Assistance]
↓
[Continue?]
↙   ↘
[Deepen Role]   [Exit / Switch]

Critical principles:
- No "loser identity"
- No permanent negation
- Exit ≠ elimination
- Only an update of mismatch information

IV. What Replaces Evaluation?

Not "no evaluation," but a three-layer feedback system

✕ **Eliminated:**
- Scores
- Rankings
- Ratings
- Pass / fail labels

✅ Retained (in a fundamentally transformed form)

① Process Feedback
- What happened?
- Where did friction occur?
- What options were available?
- Where were the risk points?

👉 Describes reality, does not judge the person.

② Self-alignment Feedback
- Am I using my real abilities?
- Do I feel energy or depletion?
- Do I want to continue?

👉 Judgment authority remains with the individual.

③ Reality Feedback
- Was the problem actually addressed?
- What impact occurred?
- Should the path change, or the actor change?

👉 Reality is the only final arbiter.

V. How Is Social Trust Built?

Not through credentials, but through Life Trajectories

The only retained "record system" in the new civilization:

🐛 Life Exploration Archive

Contains:
- Problems genuinely engaged with
- Paths attempted
- Decision rationales

- Failures and corrections
- Collaboration records
- Self-reflection traces

! Not for comparison
! Not for ranking
☑ For understanding *how* a person acts in reality

VI. How This Works at the National / Social Level (Concrete and Real)

Healthcare & Public Crisis
- Problems precede credentials
- Rapid trial-based teams
- Real-time replacement when needed

Engineering & Infrastructure
- On-site capability outweighs résumés
- Phased responsibility systems
- AI-assisted process monitoring

Technology & Innovation
- Open-source-like structures
- Whoever can solve enters
- Natural stratification without exams

Community & Public Affairs
- Problem-driven engagement
- Resident participation
- Pilot → expansion pathways

This is exactly how humanity already operates during war, disaster, and pandemics —
the difference is that it is now normalized.

VII. Why AI Is the Final Missing Piece That Makes This Civilization Possible

Because AI makes the following feasible for the first time:
- Long-term, low-cost recording of real behavior
- Non-linear capability recognition
- Dynamic matching between problems and people
- Decentralized collaboration
- Trust without exams

☞ AI does not "teach people"
☞ AI removes the structural necessity of exams

VIII. The Foundational Assumption Shift of Civilization

Old civilization assumption:

Humans are not trustworthy and must be tested, filtered, and conditioned.

New civilization assumption:

In real problems, human capability naturally reveals itself.

This is a shift in the **civilization's trust structure**.

One-sentence conclusion

At humanity's most critical moments,
we have never relied on exams to choose people.

The AI era simply allows that reality
to become the norm — for the first time in history.

Chapter 19 | Abolishing Grades, Classes, and Age: Why Life Has No Universal Timeline

Grades, classes, and age-based grouping
appear natural, rational, and unavoidable.

Yet they are among the **least questioned and most destructive institutional designs** in human society.

They take a false assumption—
one that has never existed in real life—
and hard-code it into civilization as "common sense":

> **That all human beings should grow at the same speed.**
>
> This single assumption has produced
> mass self-doubt, wasted potential, and lifelong misalignment.

I. The Real Origin of the Grade System: Management, Not Children

The grade system was never designed for children.
It was designed for **administrative survival**.

It emerged during early industrialization under conditions of:
- Rapid population concentration
- Urbanization
- Severe teacher shortages
- Limited space, time, and funding

Society needed a way to:
- Process large numbers of same-age individuals
- In fixed locations
- Within standardized schedules
- At the lowest possible cost

The grade system solved:
- Queue management
- Batch processing
- Time synchronization
- Cost reduction

It never solved:
- Understanding individual development
- Protecting cognitive diversity
- Allowing potential to unfold naturally

At its core, the grade system is **logistics logic**— not life logic.

II. Age Has Never Determined Ability, Interest, or Maturity

Reality continuously contradicts age-based assumptions:
- Some children display advanced abstract reasoning at age 6 or 7
- Others discover genuine curiosity only in their mid-teens
- Some enter deep creative phases after 30
- Many reach mastery in only one narrow domain over an entire lifetime

These are not exceptions.
They are the **normal patterns of human potential**.

What is abnormal is the system's refusal to acknowledge them.

Age is a biological marker,
yet it has been falsely repurposed as:
- A learning-capacity indicator
- A maturity metric
- A value hierarchy

This conceptual substitution has quietly governed society for over a century.

III. How Uniform Pace Creates Systemic Harm

When a system enforces a single timeline,
it does not create fairness.
It inflicts **distributed damage** on everyone.

1 Those Who Move Faster
- Forced to wait
- Curiosity eroded
- Self-direction replaced by compliance
- Creativity flattened into conformity

They do not learn exploration.
They learn restraint.

2 Those Who Move Slower
- Repeatedly labeled "behind"
- Confidence structurally eroded
- Eventually internalize: *"I am not capable"*

They do not quit learning.
They quit trusting themselves.

3 Those Who Move Differently
- Misjudged as "problematic"
- Corrected, redirected, or pathologized
- Learn to hide authentic interests and judgment

Uniform pacing does not reduce inequality.
It **equalizes harm**.

IV. The Classroom: A Long-Term Psychological Segregation Device

Classrooms appear to be learning units.
In reality, they function as **persistent identity-labeling systems**.

Once divisions appear—
- Advanced vs. regular
- Fast vs. slow
- Elite vs. ordinary

Individuals quickly internalize a belief:

"This is the kind of person I am."

This belief often outlives school itself.

Decades later, adults still limit their choices based on identities assigned in childhood classrooms.

V. AI Removes the Last Excuse for Uniform Progress

Before AI, society could claim:

"Without unified pacing, education is impossible."

That argument no longer holds.

Today:
- Knowledge is instantly accessible
- Learning pace can be fully personalized
- Paths can branch infinitely
- Depth can be entered at any moment

Maintaining grades and classes is no longer a technical necessity.
It is **institutional inertia and control habit**.

VI. Open Sessions: The Only Coherent Alternative Structure

Abolishing grades and classes does not mean chaos.
It means replacing artificial timelines with **life-aligned structures**.

Open Sessions

Core characteristics:
- No age limits
- No entry thresholds
- No completion requirements
- No evaluative outcomes

Participation is guided by:
- Current interest
- Present capacity
- Individual rhythm

People enter, exit, and return freely.

Learning is no longer assigned.
It is **permitted**.

VII. What Happens When Age No Longer Determines Position

When age is detached from status and placement:
- Intergenerational dialogue becomes normal
- Youth are no longer structurally underestimated
- Older adults are no longer structurally discarded
- Learning returns to public life

People meet not as "age categories"
but as **individuals in exploration**.

VIII. Life Is Not a Schedule—It Is an Expandable Map

The grade system is built on a silent threat:

"If you fall behind now, you will never catch up."

Reality says the opposite.

Life is not linear.
It is an expandable map:
- Direction can change
- Depth can begin at any moment
- Peaks can occur at any age
- Pauses do not equal failure

The role of institutions is not to dictate routes,
but to **avoid blocking deviation**.

Conclusion

Grades, classes, and age hierarchies
are not natural laws.
They are historical management artifacts.

Maintaining them in the AI era
is equivalent to navigating with an obsolete map
toward a destination that no longer exists.

Abolishing grades and classes does not create disorder.
It is the first act of **respect for real human tempo**.

Chapter 20 | The New Role of Schools: From Educational Institutions to Human Potential Interaction Fields

After we abolish exams, grades, classes, and age-based order,
one question naturally arises:

**If schools no longer "educate,"
what purpose do they still serve?**

The answer is not abstract.
On the contrary, it is concrete, practical, and irreplaceable.

I. Schools Were Never Just Places for Teaching

Even in the most traditional societies,
schools were never solely about the transmission of knowledge.

They have always also been:
- Spaces where people encounter one another
- Starting points of social relationships
- Places where identity is acknowledged and made visible
- Nodes of shared public resources

The issue is not whether schools matter.
The real issue is that we have long confined schools
to a function that has already become obsolete.

II. When "Education" Disappears, Spatial Value Emerges

In the age of AI, the following functions have been fully outsourced:
- Knowledge explanation
- Standardized demonstration
- Repetitive training

199

- Information delivery

Yet one category of value has become increasingly scarce in the technological era:
- Deep dialogue
- Collaborative creation
- Long-term co-presence
- Real human interaction
- Mutual understanding and trust

These cannot be replaced by online systems.

As a result, the role of schools undergoes a fundamental reversal.

III. A New Definition of School: A Human Potential Interaction Field

Within the **Human Growth** framework,
a school is redefined as:

A public infrastructure designed for human potential to meet, collide, collaborate, and unfold

It no longer operates around:
- Timetables
- Grade progression
- Instructional objectives

Instead, it revolves around one central question:
- What is happening to people right now?
- What are they exploring?
- What are they trying to solve?

IV. Schools Open to Everyone: No Age, No Threshold, No Permission

In the new civilization, schools are default public spaces.

They are not:
- Places entered at the "right age"
- Systems accessed only through qualification
- Institutions one "graduates from" after a fixed phase

They are instead:
- Open to all ages
- Open to all backgrounds
- Open to all stages of life

There are no:
- Age groupings
- Grade levels
- Admission requirements
- Completion thresholds
- Concepts of "too early" or "too late"

People no longer ask:
- "Am I suitable?"
- "Am I falling behind?"

They ask only one question:

"Is this topic real for me right now?"

V. Learning No Longer Revolves Around People, but Around Open Topics

The logic of old schools was:

Organize people first → then distribute content

The logic of new schools is:

Topics emerge first → people gather freely

Schools are formed organically around **Open Topics**.

These topics may include:
*   Technical challenges
*   Social issues
*   Community dilemmas
*   Creative themes
*   Philosophical and existential questions

For example:
*   How should cities respond to climate migration?
*   How is trust built in an AI-driven society?
*   How can local economies survive in the age of automation?
*   What is a meaningful life beyond career identity?

Anyone may participate:
*   A 14-year-old student
*   A 35-year-old professional
*   A retired engineer
*   Parents, artists, migrants, researchers
*   Individuals with no "relevant background" at all

Age is no longer relevant.
Interest and genuine motivation are the only entry points.

VI. A Fundamental Spatial Shift: From Classrooms to Fields

The core unit of traditional schools was the classroom.
The core unit of new schools is the **Field**.

These include, but are not limited to:

- **Exploration Fields**

For problem formation, discussion, trial, and failure
- **Studios / Maker Spaces**

For building, experimenting, and materializing ideas
- **AI Co-Creation Zones**

Where humans and AI think, simulate, and design together
- **Civic Labs**

Addressing real societal issues rather than hypothetical tasks
- **Reflection Spaces**

For contemplation, review, and internal integration

These spaces have no fixed ownership.
They exist only in the state of being genuinely used.

VII. The Boundary Between School and Society Disappears

In the old system, schools were seen as:

"Preparation before entering society"

In the new system, this distinction vanishes:
- Social problems enter schools directly
- School outcomes flow directly into society
- Communities, institutions, and individuals move freely in and out

Schools are no longer "before society."
They are society in motion.

VIII. What Happens When Space Stops Forcing Behavior

When schools no longer impose behavior through rules, pacing, or evaluation,
three changes naturally occur:

1 A Rise in Initiative

When choice returns to individuals,
people take responsibility for their actions.

2 Stronger Willingness to Collaborate

In non-competitive environments,
collaboration becomes a rational choice, not a moral demand.

3 A Sharp Decline in Identity Anxiety

When ranking disappears,
people exist as "what I am doing"
rather than "who I am supposed to be."

IX. Schools No Longer "Shape People," They Stop Obstructing Them

This is the most important shift.

New schools do not:
- Manufacture ideal individuals
- Export correct personalities
- Prescribe life paths

They follow one **negative yet profoundly powerful principle**:

Do not obstruct.
- Do not obstruct interests
- Do not obstruct pace
- Do not obstruct exploration
- Do not obstruct change of direction

Many human potentials do not need to be "activated."
They simply need constraints to be removed.

X. Schools as True Public Infrastructure

When schools stop producing "qualified products,"
they finally become genuine public spaces.

Just as:
- Parks do not dictate how you walk
- Libraries do not dictate what you read

Schools no longer decide:
- What kind of person you must become

They guarantee only one thing:

**When you wish to unfold yourself,
there is space, resources, and companionship
available.**

Conclusion

The school of the future
is no longer an assembly line for answers,
but a landscape where questions are allowed to grow.

It does not teach people how to succeed.
It protects one essential condition:

**When a person attempts to become themselves,
the system does not stand against them.**

This is not an upgrade of education.
It is civilization's first genuine act of respect toward
human beings.

Chapter 21 | The End of the Teacher, and the Emergence of the Guide

When those who "know" no longer stand in front,
those who are exploring can finally stand upright.

The role of the *Teacher* is not a natural feature of human society. It is not an eternal pillar of civilization, but a structure born from very specific historical conditions.

When knowledge was scarce, information was closed, and learning paths were narrow and linear,
placing the "knower" in front and the "learner" below appeared efficient—even reasonable.

In the age of AI, those conditions have collapsed entirely.

The end of the Teacher is not an ideological choice.
It is a consequence of reality already unfolding.

I. The Hidden Assumptions Behind the Teacher Have Failed

The role of Teacher rests on three implicit assumptions that were rarely questioned, yet universally enforced:

1. Knowledge is asymmetrically held
Some people know; others do not.

2. Learning paths can be standardized
Everyone should progress in the same order, at the same pace.

3. Authority is more efficient than exploration
Obedience is safer than judgment.

Today, all three assumptions fail simultaneously:
- Knowledge is instantly accessible and overwhelmingly abundant

- Learning paths are radically individual and nonlinear
- Authority increasingly obstructs judgment rather than enabling it

To preserve the Teacher under these conditions is not respect for tradition—it is denial of reality.

II. The Problem Is Not Teachers, but the Teacher Role

This must be stated clearly:
this is not a moral critique of individual teachers.

On the contrary, many educators have been forced to occupy a role that conflicts with their own judgment:
- Teaching content they know is obsolete
- Enforcing uniform pace despite obvious differences
- Preparing students for exams they recognize as harmful
- Suppressing exploration to maintain institutional order

The failure lies not with individuals, but with the structural logic of the Teacher role itself.

Any role that requires sustained suppression of human judgment is already unethical.

III. The Guide: A Fundamentally Different Role

In the **Human Growth** framework, the Teacher is replaced by an entirely different role:

The Guide.

This is not a cosmetic change—it is a civilizational shift.

- The Teacher stands *in front*
- The Guide stands *beside*
- The Teacher delivers answers
- The Guide protects the process of exploration

The Guide exists not to transmit knowledge, but to safeguard the integrity of becoming.

IV. What a Guide Explicitly Does Not Do

The Guide's first responsibility is restraint.

A Guide does **not**:
- Deliver standard answers
- Determine learning paths
- Judge correctness or hierarchy
- Set universal goals
- Declare success or failure

The moment these powers reappear, exploration collapses into compliance.

Growth cannot survive judgment.

V. The Five Core Responsibilities of a Guide

The Guide's value lies in five domains long neglected by traditional education:

1. Recognizing Signals of Potential
Not ranking ability, but noticing where energy, persistence, and genuine engagement appear.

2. Helping Form Questions Instead of Giving Answers
Answers expire.
Good questions shape a lifetime.

3. Connecting People, Resources, and Real Problems
Linking individuals with tools, AI, spaces, and societal challenges.

4. Protecting Pace and Psychological Boundaries
Shielding individuals from anxiety, comparison, and systemic pressure.

5. Accompanying Confusion Without Correction
Disorientation is not failure—it is reconfiguration.

VI. Guide and AI: Complementary, Not Competitive

In the old system, Teachers and AI inevitably conflict—both claim authority over knowledge.

In the new system, Guides and AI are structurally complementary:
- **AI** handles knowledge, simulation, and possibility expansion
- **Guides** handle meaning, judgment, ethics, and direction

AI can answer *how*.
Only humans can help navigate *why* and *who*.

VII. Why Guides Are More Scarce Than Teachers

This is the paradox of the AI age.

Before AI, society needed people who could explain clearly.
After AI, society needs people who can:
- Tolerate uncertainty
- Respect difference
- Resist control
- Remain present without domination

These capacities are far rarer than subject expertise.

VIII. The Ethical Boundary of the Guide

The Guide's highest ethical principle is simple:

I do not decide who you should become.

> The moment a Guide shapes identity, they revert to authority.
> The moment they abandon judgment entirely, they become spectators.
>
> The Guide's position lies precisely between these extremes.

Conclusion

The end of the Teacher is not the collapse of education.
It is the maturation of civilization.

When a society no longer requires
those who "know more"
to dominate those who are becoming,

when authority no longer overrides judgment,

only then does society truly begin to trust human beings.

Chapter 22 | Stop Educating Your Children

Rewriting the Role of Parents in the Age of Human Growth

Almost every parent says the same sentence:

"Everything I do is for my child's own good."

It sounds unquestionable.
Yet this sentence forms the most stable—and most dangerous—
emotional foundation of the old education civilization.

Because it allows structural control
to exist indefinitely
under the name of love.

From the perspective of **Human Growth**, we must confront an
uncomfortable but unavoidable truth:

**Much of what is called "education" is not a condition for
growth—
it is an obstruction to it.**

I. Parents Are Not the Problem — "Educational Love" Is

First, something must be made clear:

Parental love is real.

The problem is not love itself,
but that this love has been captured by a flawed civilizational
logic—
the logic of education.

In the old system, loving a child was gradually redefined as:
- Designing their life path
- Eliminating uncertainty on their behalf

- Placing them into an approved position called "success"

From this emerged a specific form of love:

Educational Love

It consistently shows three characteristics:

1 Deciding the child's direction for them
Uncertainty is treated as danger; choice as risk.

2 Using the future to deny the present
Current interests, joy, and inner states are sacrificed for "what comes later."

3 Packaging anxiety as responsibility
Control becomes care. Interference becomes protection.

This is not cruelty.
It is a conditioned response trained by the system.

II. Why Parents Find It So Hard to Let Go

In the old civilization, a child's life often becomes
the last future parents believe they can still control.
- Society feels uncontrollable
- The economy feels uncontrollable
- Technology feels uncontrollable
- The world feels uncontrollable

So the child becomes the final territory left to design.

Education turns into an emotional compensation mechanism.

You think you are shaping your child,

but often you are trying to prove:

"At least here, I still have control."

III. What You Are Educating Is Often Your Fear of Failure

This is a harsh but honest truth:

**Many educational actions are not for the child,
but to help parents avoid facing their own fears.**

These fears include:
- Fear the child will "fall behind"
- Fear of being judged as irresponsible
- Fear the child will suffer in the future
- Fear that one's own choices will be proven wrong

So, under the banner of love:
- Interests are suppressed
- Natural rhythms are overridden
- Directions are locked in too early

Control justified by love is still control.

And people who grow up under control
rarely develop real judgment or responsibility.

IV. Human Growth Requires a New Parental Identity

When "education" exits the center of civilization,
parents must undergo a fundamental identity shift.

213

From:

Life Designers

To:

Guardians of Growth

This is not neglect.
It is a far more demanding responsibility.

V. What Parents Do in a Human Growth Framework

In **Human Growth**, parents are not responsible for outcomes.
They are responsible for **protecting the conditions of growth**.

A growth-oriented parent does only four things:

1 They do not decide the child's direction

Direction must come from within.
Otherwise, collapse is only delayed.

Parents may share experience and reality—
but they do not walk the path for the child.

2 They do not use comparison as motivation

Comparison produces only:
  •   Anxiety
  •   Imitation
  •   Self-negation

It never creates genuine inner drive.

3 They do not use the future to invalidate the present

A child is not an "unfinished person."
They are already whole—living a real stage of life.

In Human Growth:

The present is not a tool.
It is life itself.

4 They do not rush to fix confusion

Confusion is not failure.
It is a signal that growth is happening.

Immediate correction interrupts exploration.

VI. The Hardest Lesson: Allowing Failure

In Human Growth:

Failure is not an accident.
It is part of the path.

But this demands something extremely difficult from parents:

Accepting that the child may not succeed
in the way you imagined.

This is real love—
because it requires letting go of:
- Control
- Projection
- The fantasy of a "correct life"

Only this kind of love prevents a child's life
from becoming an extension of parental fear.

VII. What Happens When Parents Stop Educating

The outcome is often unexpected:
- Conflict decreases
- Trust increases
- Conversations become real

Because the child no longer needs to perform, please, or prove worthiness.

They can finally live—
not be managed.

VIII. The Ultimate Responsibility of a Parent

It is not delivering the child to a destination.

It is answering one question:

When you are no longer responsible for their life, can they take responsibility for themselves?

If the answer is no,
then even the most "successful" education
is only delayed failure.

Conclusion | Give Life Back to the Child

Stop educating your children.

If you truly love them,
return their life to them.

Human Growth is not abandonment.
It is civilization's first real acknowledgment that:

**A child is not a parental project,
but a life in the process of becoming.**

Chapter 23 | A New Definition of Learning

(And the End of the Old One)

When School and Society Reunite, Learning Becomes Real Life

Learning is no longer the acquisition of knowledge.

It is:

the lifelong process through which a person unfolds themselves,
understands the world,
builds relationships,
and gradually learns to carry responsibility.

It is:
- lifelong
- non-competitive
- non-rankable
- non-quantifiable

And precisely because of that,
it is the only form of learning that truly matters.

I. The Most Dangerous Fiction: "First Learn, Then Live"

In traditional systems, school and society are artificially split into two stages:
- first, **learning**
- later, **life**

As if a human life could be divided into:
- a preparation phase
- and a real beginning

This is one of the most damaging constructions of industrial civilization.

Because it traps entire generations in a suspended psychological state:

"My real life hasn't started yet."

People grow up rehearsing existence
instead of inhabiting it.

II. "Preparing for the Future" — A Narrative Used to Justify Control

Nearly every system that suppresses present freedom relies on the same sentence:

"This is for your future."

The problem is not intention.
The problem is that this narrative creates a permanently postponed life.

Under its logic:
- interests are delayed
- judgment is suspended
- choices are replaced
- responsibility is removed

When people are repeatedly told
that they are "not yet ready" to face real problems,
they eventually lose the ability
to face real problems at all.

III. Learning Decays Rapidly When Removed from Reality

The moment learning is sealed inside institutions,
three distortions inevitably occur:

1 Problems become artificial

Educational problems are designed to be solved
without real consequences.

They train performance, not judgment.

2 Outcomes become performances

Assignments exist to be evaluated,
not to create impact.

Success becomes appearance,
not effect.

3 Responsibility is absorbed by the system

Failure is reduced to a score,
not a lived consequence.

This is not learning.
It is a simulation of life,
carefully designed to feel safe.

IV. The Real World Is the Only Legitimate Source of Learning

Within the **Human Growth** framework,
learning no longer begins with a curriculum.

It begins with:

- a real problem
- a real need
- a real conflict
- a real unknown

Learning is no longer organized around disciplines,
but around situations:
- environmental crises
- community governance
- technological ethics
- cultural expression
- human relationships

These problems have no standard answers.
They require judgment, not recall.

V. What Happens When Schools Open Themselves to Society

The moment schools stop treating themselves as
"what comes before society"
and recognize themselves as part of society,
several changes occur immediately:

1 Learning regains weight

Actions now have consequences.

2 Motivation becomes intrinsic

People stop asking,
"Why do I need to learn this?"
because the problem is already present.

3 Responsibility returns

Choices are no longer exercises.
They are interventions in reality.

VI. Learning Outcomes Are No Longer "Submitted" — They Enter the World

In the new system, there is no such thing as "handing in homework."

Learning outcomes go to only three places:
- into communities
- into public discussion
- into real-world use

They may take the form of:
- a solution
- an action
- a tool
- an expression
- or a correction after failure

The world itself becomes the feedback system.

VII. Failure Reclaims Its True Meaning

In school-society separation systems, failure is controlled and isolated:
- it exists only on paper
- its consequences are softened
- responsibility is abstracted

In real contexts, failure regains its original function:

Failure is not punishment.
It is a method of understanding reality.

When failure is no longer graded,
it becomes meaningful for the first time—
even though we no longer call it "education."

VIII. Children and Adults Learn in the Same Reality

When learning re-enters society,
age boundaries dissolve naturally:
- children participate in real projects
- adults return to being learners
- experience no longer silences questions
- authority is no longer monopolized by age

People meet in a more honest way:

Not as "those who teach" and "those who are taught,"
but as people facing the same problems together.

IX. Society No Longer "Uses" Schools — It Participates in Them

In the old system, society was merely the destination after school.

In the new system, society becomes part of learning itself:
- communities enter learning spaces
- public issues become shared inquiries
- conflicts become discussion material
- uncertainty becomes acceptable

This does not create chaos.

It creates alignment.

X. When Learning Becomes Life, Life Is No Longer Divided

When people are no longer told
that they are in a "preparation phase,"
they realize something fundamental:

This moment is life.

Choices carry weight.
Actions matter.
Time is no longer spent waiting for permission.

Conclusion | Learning Returns to Where It Belongs

Separating learning from life
was the price industrial civilization paid
to manage human beings efficiently.

In the age of AI,
that price has become unbearable.

When learning returns to the real world,
life finally begins.

And for the first time,
**Human Growth becomes possible—not as a theory,
but as a lived reality.**

Chapter 24 | A New Word After "Education": The Human Growth Center

We refuse to continue using the word **"education."**

Not because learning is no longer important,
but because this word itself can no longer describe reality.

"Education" carries with it an entire set of civilizational
assumptions that have already failed:
- Humans are objects to be shaped
- Growth requires external authority
- Value can be uniformly measured
- Order must be maintained through obedience

These assumptions belong to **industrial civilization**,
not to the **age of AI**.

Therefore, we introduce a new core concept:

The Human Growth Center

Humans are not cultivated.
They are not molded.
They are not produced.

Humans grow.

Schools do not create people.
They can only do one of two things:

Support growth — or obstruct it.

Human Growth Center

After Education in the Age of AI

I. Education Has Completed Its Historical Mission

Modern society treats "education" as an eternally correct institution,
as if wherever society exists, education must exist in its current form.

This is a profound historical illusion.

Education is not a fundamental structure of human civilization.
It is a **stage-specific population management tool**,
designed to organize human growth
at the lowest possible cost under specific technological conditions.

When those conditions change fundamentally,
the tool should be set aside —
not preserved as sacred tradition.

Today, we stand precisely at such a historical moment.

II. Education Was Never Designed for People to Become Themselves

Historically, modern education systems took shape in two major transitions:
- From agrarian societies to feudal orders
- From feudal orders to industrial civilization

In both cases, the core objective of education was never human fulfillment or complete growth.
It was:
- Stable transmission
- Behavioral uniformity
- Obedience to order
- Predictability

225

Industrial education, in particular, demanded not unique individuals,
but **replaceable, manageable, rankable people**.

Grades, classes, schedules, exams, certificates —
none of these were designed to understand individual differences.
They were designed to **assemble humans into systems**.

III. Exams: A Control Technology Mistaken for Fairness

Exams have long been packaged as "objective," "rational," and "fair."
But they have never measured intelligence.

They measure only three things:

1.   Obedience to standard answers
2.   Ability to function under pressure
3.   Adaptation to system rules

This made sense in the industrial era,
when machines required stable, compliant humans.

But in the age of AI:
* These abilities are no longer scarce
* They no longer define social value
* They no longer ground human dignity

Continuing to define people through exams
produces nothing but **systemic failure and internalized self-denial**.

IV. AI Has Broken the Logic of "Learning" Itself

AI does not represent an upgrade to educational technology.
It represents a fundamental negation of the question:

Why does education exist at all?

When a child can freely discuss with AI:
- Quantum physics
- Ethical philosophy
- Art history
- Engineering design

we must confront an undeniable fact:

Humans no longer need to be "taught" in order to access knowledge.

Maintaining a knowledge-delivery-centered education system today
is like teaching people to repair telegraphs in the smartphone era.

V. What Has Become Truly Scarce Is the Human Being

Once knowledge, skills, and memory are fully covered by AI,
humanity is forced to face a long-hidden question:

If value is no longer defined by "knowing more than others,"
what is a human being worth?

The answer is unsettling — yet unmistakably clear:
- Judgment
- Sense of direction
- Sensitivity to meaning
- Understanding of others
- Awareness of one's own potential

These capacities:

- Cannot be injected
- Cannot be tested
- Cannot be standardized

Yet they determine the quality of an entire human life.

VI. Education Has Become the Ox Cart in the Age of Aircraft

A dangerous inversion now defines modern civilization:
- Technology accelerates
- Institutions stagnate
- Human suffering expands

Children are forced to compete inside failing systems.
Adults are forced to explain their lives through scores.
Societies pay continuously for structural waste.

This is not a national problem.
It is a **civilizational phase mismatch**.

VII. We Are Not Reforming Education — We Are Ending It

This is the most easily misunderstood claim of this book.

Ending education does not mean ending learning.
It means ending:
- Coercion
- Ranking
- Uniform pathways
- "Correct life templates"

And returning — for the first time —
the right to **become oneself** fully to the individual.

VIII. When Old Words Fail, New Structures Are Required

The word "education" assumes:
- Humans must be shaped
- Growth requires authority
- Value can be judged

All of these assumptions have collapsed.

Therefore, we propose a new institutional form:

The Human Growth Center

IX. What Is a Human Growth Center?

A Human Growth Center is not a school.

It is a form of **public infrastructure** governed by a single principle:

Do not obstruct human growth.

It does not:
- Define life paths
- Certify human worth
- Rank or filter people
- Produce "qualified outputs"

It provides:
- Open access for all ages
- Real problems and real contexts
- AI-assisted collaboration
- Cross-generational, cross-background participation
- Time, safety, and dignity

Growth is not manufactured there.
It is **allowed to occur**.

X. From Education to Civilizational Maturity

Education was never the destination of civilization.
It was a transitional management tool.

AI marks the end of that phase.

What humanity must now learn is not more information,
but something far more difficult:

How not to obstruct one another from becoming ourselves.

Conclusion

Education was a phase —
not a destiny.

AI has brought that phase to a close.

The **Human Growth Center** is not an upgraded school system.
It is **a post-education civilizational structure**.

This is not chaos.
It is humanity's first real step
into a mature society built on **trust, reality, and responsibility**.

Final Chapter | The AI Era: Humanity's First Global Equalization of Human Growth

The First Global Equalization of Human Growth

I. AI Does Not Bring "Better Education" — It Brings the First True Equal Starting Point

Throughout human history,
place of birth has largely determined the upper limit of a life.

No matter how much education systems were improved,
one brutal structural reality never changed:
- Education depends heavily on national wealth
- It requires vast teaching labor forces
- It relies on administrative capacity and long-term institutional stability

Traditional education therefore carries an unavoidable structural outcome:

Rich countries can keep optimizing.
Poor countries can only keep chasing.

AI breaks this chain for the first time.

When cognition, language, reasoning, and understanding
can be accessed globally on the **same technological platform**,
humanity stands—*for the first time*—at a **shared starting point for human growth**.

II. One Platform, One Capability Level — A First Alignment in Civilizational History

AI does not bring the world into "the same curriculum."
It brings the world into **the same capability layer**.

This means:
- A child in a rural African village
- A teenager in a South Asian slum
- A young person in a marginalized Latin American community

can, with AI collaboration:
- Use cognitive tools at the *same level* as those in developed countries
- Enter equally deep structures of understanding and reasoning
- Engage directly with global knowledge, science, art, and ethics

For the first time in human history,
the ability to understand the world is no longer priced by nationality, family, or class.

III. The Cost Revolution: AI Is Not More Expensive — It Is Radically Efficient

Traditional education systems are, by nature:
- Labor-intensive
- Administration-heavy
- Poorly scalable
- Highly repetitive in construction

The **AI + Human Growth** model does not merely "save some budget."
It causes a **collapse in cost by entire orders of magnitude**.

The structural shift is fundamental:
- No need for massive teaching workforces
- No need for layered education bureaucracies
- No need for exams, rankings, certifications, or credential systems

• No need to endlessly replicate standardized schools

For the first time,
providing growth conditions for everyone becomes economically realistic.

IV. The Real Path for Poor Countries: Human Growth Centers, Not School Replication

For low-income countries,
the truly unrealistic path has never been "education reform."

It has always been the attempt to **replicate rich-country school systems**.

This framework proposes a fundamentally different—and more humane—path:

Human Growth Center

Not a school,
but a **minimal-institution, maximum-efficiency human infrastructure**.

Core Characteristics
• Fully open to all children
• No grades, no exams, no diplomas
• AI as the primary cognitive collaboration system
• A very small number of Guides focused on companionship and ethical safeguarding

V. Learning and Survival Are Solved Together — This Is the Key Difference

In many poor regions,
the real reason children cannot grow is not "lack of learning,"
but:
- Hunger
- Insecurity
- Unsafe living conditions
- Family environments unable to provide stability

Human Growth Centers can simultaneously provide:
- Basic and stable nutrition
- Safe, sustainable living spaces
- Simple but healthy dormitories
- Living conditions clearly better than surrounding environments

And even with these provisions,
the **total cost remains far below that of traditional education systems**.

This is not an added benefit.
It is a direct response to **why learning fails in reality**.

VI. One Design, Coverage Everywhere

Because this system:
- Does not depend on teacher scale
- Requires no administrative bureaucracy
- Has no hierarchical management layers
- Needs no standardized evaluation mechanisms

It achieves something traditional education never could:

One design that works everywhere.

Cities, villages, remote regions, refugee zones.

Wherever there is:
- • Basic connectivity
- • Minimal energy
- • A secure physical space

the same Human Growth platform can operate.

VII. This Is Not Charity — It Is a Reconfiguration of Civilizational Efficiency

This must be stated clearly:

This is not about "helping the poor learn."

It is humanity's first collective decision
to stop systematically wasting human potential.

The greatest hidden cost of traditional education was never financial.
It was:
- • Abandoned talent
- • Suppressed creativity
- • Lives prematurely labeled as "failures"

The **AI × Human Growth** system is a civilizational structure that releases
the maximum human potential at the lowest institutional cost.

VIII. The True Meaning of Global Equalization

Equalization does not mean forcing everyone onto the same path.

It means:

That **any human being**,
regardless of where they are born,

235

has an equal opportunity
to grow into their **own unique humanity and potential**.

This is a mission traditional education systems could **never**
fulfill structurally.

Final Conclusion

AI makes it possible—for the first time in history—
to provide equal conditions for human growth **without relying
on national wealth**
and **without replicating expensive education bureaucracies**.

This is not merely technological progress.

It is the first true sign
that human civilization is entering maturity.

Appendix

Human Growth Manifesto

Beyond Education, Toward a New Human Society

Preamble | The Age of Education Has Completed Its Historical Mission

For most of human civilization,
education has been regarded as the central mechanism for
shaping individuals, transmitting knowledge, and maintaining
social order.

In earlier eras,
education was indeed necessary.

But **necessity does not equal permanence**.

In the age of artificial intelligence,
the foundational assumptions that sustained "education" are
collapsing systematically:
- Knowledge is no longer scarce
- Skills no longer require long-term institutional training
- Intelligence is no longer monopolized by teachers or schools
- Learning is no longer constrained by national wealth or administrative capacity

Structures that once propelled civilization forward
are now becoming barriers to human development.

Education is no longer the engine of human progress.
It is increasingly an institutional legacy of a past era.

Article I | Humans Are Not Objects to Be Shaped

Human beings are not raw materials.
They are not objects to be molded, optimized, or standardized.

Every person is born with:
- intrinsic curiosity
- adaptive intelligence
- moral intuition
- the capacity to sense meaning and care for others

Educational systems long assumed that
humans must first be "formed" in order to function.

The AI era reveals a deeper truth:

Humans do not need to be shaped.
They need to grow without being obstructed.

Article II | AI Ends the Scarcity of Education

Education existed because knowledge, reasoning, and expertise
were once scarce.

AI has permanently altered this condition.

For the first time in human history:
- explanation
- reasoning
- simulation
- translation
- reflection

are available globally, on demand,
at near-zero marginal cost.

When **intelligence itself becomes infrastructure**,

education loses its institutional monopoly.

Article III | Human Growth Replaces Education

We define **Human Growth** as:

> The natural unfolding of an individual's cognitive, ethical, creative,
> and social capacities
> when obstacles are removed and conditions are aligned.
>
> Human growth does **not** depend on:
> - classrooms
> - curricula
> - examinations
> - grade levels
> - certifications

Human growth **requires**:
- access
- safety
- time
- dignity
- cognitive tools
- ethical space

Article IV | Human Growth Centers as the Infrastructure of a New Civilization

A **Human Growth Center** is not an instructional institution.

It is a **public infrastructure of possibility**.

A Human Growth Center:
- does not teach
- does not rank

- does not evaluate
- does not determine life paths

It ensures only one thing:

That human growth is not blocked.

Within it, AI functions as:
- a cognitive amplifier
- an exploratory partner
- a reflective mirror

Never as an authority.

Article V | The First Real Condition for Global Equality

For most of history,
a child's potential was almost entirely determined by birthplace.

Education systems reinforced this inequality because they
depended on:
- national budgets
- teacher availability
- institutional stability

Human Growth Centers break this dependency.

With minimal infrastructure:
- connectivity
- energy
- safe physical space

**any child, anywhere, can access the same depth of cognition
and the same growth tools.**

For the first time in history,
human potential is no longer priced by geography.

Article VI | Cost Revolution and Civilizational Efficiency

Traditional education systems are:
- labor-intensive
- administratively heavy
- structurally repetitive
- inherently unequal

Human Growth Centers eliminate:
- teacher hierarchies
- bureaucratic layers
- standardized evaluation systems
- repetitive institutional costs

Even when combined with:
- basic nutrition
- living space
- simple dormitories

their total cost remains **significantly lower** than conventional schools.

This is not austerity.
It is civilizational efficiency.

Article VII | The Role of the Guide Is Protection, Not Direction

In a human growth society,
the central human role is no longer the Teacher,
but the **Guide**.

A Guide:
- does not instruct
- does not evaluate
- does not determine outcomes

A Guide safeguards:

- autonomy
- dignity
- ethical boundaries
- psychological safety

The Guide exists to prevent systems
from replacing human agency.

Article VIII | A New Definition of Happiness

Happiness is not:
- success
- status
- wealth
- comparison

Happiness is:

The inner coherence that emerges
when a person fully realizes their unique capacities
and takes responsibility for their own existence.

Human growth is not about producing outcomes.
It is about completing lives.

Article IX | The Minimal Ethical Principle of Civilization: Do Not Obstruct

A mature civilization requires only one foundational rule:

Do not obstruct human growth.
- Do not obstruct difference
- Do not obstruct timing
- Do not obstruct failure
- Do not obstruct exploration
- Do not obstruct becoming

This is not permissiveness.
It is restraint born of wisdom.

Final Declaration

Education was a necessary phase in human history.
But it is not the future.

The future belongs to societies that:
- stop shaping humans
- and start supporting human growth

This manifesto marks a civilizational transition:
- from **Education → Human Growth**
- from **Instruction → Conditions**
- from **Control → Dignity**

Closing

When society no longer asks
"how should we educate people,"
but instead asks
"how can we avoid obstructing human growth,"

a new human era truly begins.

Case Study | Why Poor Countries Should Skip Traditional Education and Enter the Human Growth Center Era

From Education Catch-Up to Human Growth Leapfrogging

I. Core Judgment

For poor countries, copying traditional education systems is not a development path—it is a **structural trap**.
The **Human Growth Center** is the only option that is sustainable economically, socially, and civilizationally.

II. Why Traditional Education Is the Wrong System for Poor Countries

1 **Traditional education is a high-institution-cost system that poor countries cannot sustain**

Traditional education depends on:
- Large-scale teacher workforces
- Heavy administrative structures (ministries, districts, exam boards)
- Long-term fiscal stability
- Social trust in diplomas and credentials

☞ **These conditions simply do not exist** in most poor countries.

The outcomes are almost always one of the following:
1. Extremely low education quality
2. Corruption, formalism, and fake credentials
3. Youth unemployment combined with diploma inflation

Education does not produce capability—it produces frustration.

2 "Exams + diplomas" are already dysfunctional in poor societies

Before the AI era, diplomas still had some filtering function.
After AI, diplomas in poor countries are effectively bankrupt:
- Local economies cannot absorb graduates
- Diplomas no longer signal real ability
- Youth are trained to *wait for jobs*, not to create paths

☞ In poor countries, exam systems now serve one real function:
They delay failure instead of creating a future.

3 What poor countries lack is not knowledge, but independent thinking and action

Young people in poor countries are not lacking:
- Motivation to learn
- Real-world problems
- Survival drive

What they lack is:
- Independent judgment
- Problem-modeling ability
- Self-directed growth pathways

Traditional education systematically suppresses all three.

III. Why Human Growth Centers Are Naturally Compatible with Poor Countries

Core Logic

Human Growth Centers do **not** require the state to be strong first. They only require **not blocking human growth**.

1 AI directly eliminates the teacher gap

Human Growth Centers do not require:
- Good teachers
- Standardized textbooks
- Unified curricula

AI provides:
- Explanation
- Reasoning and simulation
- Multilingual support
- Cognitive companionship

☞ This is the first time poor countries can access **global-level cognitive tools** directly.

2 No exams makes Human Growth more suitable—not less

Removing exams and certificates leads to:
- Youth no longer waiting for recognition
- Learning driven by reality, not credentials
- Earlier engagement with real problems
- Earlier formation of responsibility

For poor countries, **no certificates are not a loss—they are liberation**.

—

3 Human Growth Centers address learning and survival simultaneously

In poor regions, the main reason learning fails is not laziness—it is survival pressure.

Human Growth Centers can integrate:
- Basic food security
- Safe physical space
- Simple dormitories
- Internet and AI access

☞ Even with these services included, total costs remain far lower than traditional schools—while results are more real.

IV. Prototype Model | Human Growth Center (Poor-Country Version)

Positioning
- District-level
- Youth aged 12–25
- Open daily
- No enrollment, no grades, no graduation

1 Coverage Capacity (per center)

Item	Scale
Resident youth	300
Daily open participation	500–800
Annual youth reached	1,500–2,000

2 Spatial Design (Minimalist)
- Open learning space (AI access)
- Quiet thinking areas

247

- Collaboration / project zones
- Basic dormitories (150–200 beds)
- Communal food area
- Basic sanitation

No:

- Classrooms
- Administrative offices
- Teacher lounges
- Bureaucratic buildings

3 Staffing Structure (Ultra-Light)

Role	Number	Description
Guides	6–8	Not teachers; safety & ethics only
Tech support	2	Network & devices
Logistics	4–6	Food & hygiene
Coordination	1	Minimal management

☞ **No teachers. No education bureaucracy.**

4 Cost Structure (Annual Estimate)

Item	Annual Cost (USD)
Facility & maintenance	$80,000
Internet & AI access	$40,000
Food support	$120,000
Dormitory operation	$60,000
Personnel	$150,000
Total	≈ $450,000 / year

Per-capita annual cost
- Based on 1,500 youth/year
☞ **≈ $300 per person per year**

This is **20–40% of traditional school costs**,
while delivering far more authentic growth capacity.

V. Social Outcomes (3–5 Years)

Human Growth Centers do not promise "employment rates."
They produce **structural change**:
- Earlier independent thinking
- Stronger problem awareness
- Deeper community participation
- Higher entrepreneurship and self-organization
- Fewer "credential-deceived generations"

Society no longer waits for a "well-educated generation."
It begins to see **a generation that is already growing**.

VI. Final Conclusion

Poor countries should not waste scarce resources copying
an education system already obsolete in the AI era.

Human Growth Centers allow them to:
- Use the lowest possible cost
- Cultivate the most independent and creative generation
- Leap directly into a post-education civilization

This is not catching up.

This is leapfrogging.

Global Guide Selection and Development Framework

A Civilizational Role for the Age Beyond Education

I. The Civilizational Position of the Guide

A **Guide** is **not**:
- a teacher
- a mentor
- a counselor
- a psychotherapist
- a life coach

So what *is* a Guide?

A **Guide is a civilizational role**.

In a **post-education society**, the Guide exists for one purpose only:

> **To prevent systems from flattening human beings again.**

The Guide's sole reason for existence is this:

> **When a person is unfolding their own life, someone must ensure that no one else decides for them.**

II. Why Guides Cannot Be Replaced by AI or Institutions

Because the Guide's core capacities are not knowledge-based.

They are the ability to:
- tolerate uncertainty
- respect difference
- restrain power
- protect rhythm
- coexist with situations that have no answer

These capacities:
- cannot be quantified
- cannot be graded
- cannot be automated

They are **human limits**, not technical skills.

III. The Global Guide Competency Model

(Five Non-Negotiable Core Capacities)

These five capacities form the **minimum global standard**, independent of culture, nation, or ideology.

1 Tolerance for Uncertainty
- Not rushing to conclusions
- Not fearing ambiguity
- Not using "efficiency" to crush exploration

☞ This is the primary dividing line between **control-oriented personalities** and true Guides.

251

2 Non-Shaping Presence
- Not implying a "better life template"
- Not projecting personal values as direction
- Not influencing through "I'm doing this for your own good"

☞ The moment a Guide tries to *shape*, the role collapses.

3 Deep Listening and Inquiry
- Hearing what is not yet spoken
- Asking questions without steering answers
- Being able to remain present in long silence

4 Radical Respect for Difference
- No judgment based on culture, class, age, or personality
- No anxiety toward slowness, detours, or repetition

5 Self-Awareness and Power Restraint
- Recognizing the impulse to rescue
- Detecting the desire to feel "useful"
- Being able to step back deliberately

IV. Global Guide Selection System

(Four-Stage Process)

Stage One | Open Application + Personal Statements (Global Standard)

Applicants **do not submit**:
- degrees
- certificates
- professional titles

They submit **only three items**:

1 A real personal failure
2 One experience of accompanying someone without answers
3 A life choice they do not understand—but fully respect

☞ This stage filters out approximately **70% of control-oriented personalities**.

Stage Two | Observation Period (3–6 Months)

Role: Observer

Observers are **not allowed to guide anyone**.

They may only:
- observe interactions
- record internal reactions
- keep reflective journals

Evaluation focuses not on *what they did*, but on:
- urgency to intervene
- discomfort with silence
- desire to prove usefulness

Stage Three | Accompaniment Period (6–12 Months)

Role: Apprentice Guide
- Shadow experienced Guides
- Permitted actions:
  - presence
  - observation
  - reflection

Strictly prohibited:
- giving advice
- setting directions
- summarizing "correct paths"

Stage Four | Ethical Confirmation (Global Standard)

There is only **one core question**:

> *"If a person ultimately chooses a life path you deeply disagree with—*
> *even one you believe is unwise—*
> *are you still willing to remain present?"*

This is **not** judged by words,
but confirmed through **long-term behavioral observation**.

—

V. Guide Development System

(Not Training, but De-Powering)

Module One | Human Growth and Disorientation Structures
- Why disorientation is a necessary phase
- Why premature clarity is a danger signal

Module Two | Dialogue Ethics (Global Red Lines)
- No 诱导 (no subtle steering)
- No replacement of judgment
- No dependency creation

Module Three | AI Collaboration Literacy
- AI is an amplifier, not a judge
- Guides do not compete with AI on intelligence
- Guides protect meaning and ethical boundaries

Module Four | Self-Awareness Training (Critical Core)
- Control impulse recognition
- Savior complex detection
- Achievement projection awareness

☞ This is the **most important** part of Guide development.

VI. Three Absolute Global Red Lines for Guides

✖ Defining success
✖ Suggesting life templates
✖ Making decisions for others

Any violation **automatically triggers review and suspension**.

VII. Guide Recognition System

(Anti-Certification Model)

✖ No levels
✖ No rankings
✖ No star Guides

There is only one status:

Active Guide

The sole condition for maintaining this status:

The ability to step back.

If a Guide:
- begins giving frequent answers
- becomes relied upon
- starts enjoying authority

☞ Immediate pause or exit is mandatory.

VIII. Why This Framework Is Globally Replicable

Because it does **not** rely on:
- cultural export
- disciplinary standards
- language dominance

It relies on one universal human truth:

> **When humans are growing, what they need most is not guidance—**
> **but the absence of interference.**

IX. The Ultimate One-Sentence Definition

> **The Guide's responsibility is not to lead people somewhere,**
> **but to ensure the world does not decide their direction for them.**

Guide International Charter

Global Charter for Guides of Human growth

Positioning:
A civilizational public role framework for the **post-education era**

Part I

Article 1 | Purpose

In the age of artificial intelligence—
when knowledge is no longer scarce,
paths are no longer singular,
and traditional educational structures are steadily losing relevance—
this Charter establishes a new public role:

Guide —
a civilizational role that does not shape people,
does not select people,
and does not optimize people,
but exists solely to **safeguard the freedom of human unfolding**.

Article 2 | Foundational Principles

1 Human dignity is not measurable
No form of ranking, evaluation, or capability quantification may serve as a basis for a Guide's actions.

2 Unfolding precedes efficiency
The rhythm of human growth must not be compressed by institutional, economic, or technological efficiency demands.

3 Difference does not require correction
Differences in culture, personality, life path, and tempo are legitimate forms of human unfolding.

4 The Guide's competence is the ability to step back
A Guide's professionalism is demonstrated precisely by non-intervention, non-replacement, and non-direction at decisive moments.

Article 3 | Role Boundaries

A Guide must not:
- define a "successful life"
- recommend specific life paths
- impose direction under the guise of goodwill
- create dependency
- act as evaluator or judge

A Guide is permitted only to:

During a person's unfolding process,
prevent external systems and invisible power structures
from making decisions on that person's behalf.

Article 4 | Non-Negotiable Ethical Red Lines

Any Guide exhibiting the following behaviors

must immediately enter a suspension process:
- influencing choices through authority-based language
- presenting personal values as a "more mature solution"
- experiencing anxiety when others move slowly or deviate
- deriving personal fulfillment from being depended upon

Article 5 | Global Certification Logic

(Anti-Credential Model)
- No ranks
- No certificates
- No international rankings

The only legitimate status is:

Active Guide

Validity depends solely on one condition:
whether the Guide still possesses the **capacity for power restraint and withdrawal**.

Article 6 | Statement of International Applicability

This Charter exports no cultural model,
no educational system,
and no ideology.

It establishes only one minimum, cross-civilizational ethical consensus:

In the AI era,
every society must preserve **institutional space
for individuals to become themselves**.

Article 7 | Civilizational Commitment

Institutions that adopt or endorse this Charter
acknowledge the following truth:

The quality of future civilization
will no longer be measured by how many "successful
individuals" it produces,
but by how many people are **not systemically suppressed**.

Part II

Global Seed Network for Human Unfolding Guides

(50–100 Individuals)

Objective

Within 24 months,
to form a cross-cultural, cross-national, cross-background
Guide seed network
serving as a **living prototype** for global replication and
local adaptation.

I. Why a "Seed Network" Is Necessary

The Guide role cannot be scaled through training programs.
Otherwise, it will immediately degenerate into:
  • a certification industry

- a consulting profession
- a new authority structure

Therefore, global propagation must follow:

An **extremely slow, highly restrained, high-ethics seed replication model**.

II. Composition Principles (50–100 People)

Required diversity dimensions:
- **Regions:** North America / Europe / East Asia / South Asia / Latin America / Africa
- **Age:** 25–70
- **Backgrounds:** Arts / Engineering / Community work / Healthcare / Public affairs / Former teachers / Former corporate managers

The only commonality:
A demonstrated withdrawal from the impulse to shape others.

III. Global Selection Path (Decentralized)

① **Recommendation-only (No open recruitment)**
- Trusted network referrals only
- Prevents self-projecting applicants

② **Long-term observation (6–12 months)**
- No titles granted
- No identities assigned
- Behavior observed only in real situations

IV. Three Core Responsibilities of Seed Guides

1 Do not replicate yourself
Any attempt to "train more people like me" triggers immediate suspension.

2 Do not export standardized language
Share experiences only.
Never summarize methods.

3 Generate locally, never expand globally
Each Guide is responsible only for a small, concrete reality.

V. Network Operations (Minimalist by Design)
- No headquarters
- No secretariat
- No KPIs
- No annual growth targets

Only three functions exist:
- Annual rotating in-person cross-cultural dialogues
- Small-scale ethical retrospectives
- Failure case sharing (non-public)

VI. Network Dissolution Mechanism (Critical)

The network must actively contract or dissolve if any of the following appear:
- Converging language
- Emergence of "star Guides"
- Market or commercial absorption
- Governmental use as a soft governance tool

☞ **Self-dissolution is one of the network's most important capabilities.**

VII. One Sentence for All Participants

You are not building an organization.
You are preventing a new organization
from once again replacing human beings.

Final Summary

Guides are not experts of future education.
They are the people who preserve, within future civilization,
the final institutional boundary ensuring that:

Human beings remain ends, not means.

www.ingramcontent.com/pod-product-compliance
Lightning Source LLC
Chambersburg PA
CBHW071538200326
41519CB00021BB/6536